MANAGING
HIGH-TECHNOLOGY
PROGRAMS
AND
PROJECTS

MANAGING HIGH-TECHNOLOGY PROGRAMS

AND

PROJECTS

2nd Edition

Russell D. Archibald

John Wiley & Sons, Inc.
New York • Chichester • Brisbane • Toronto • Singapore

This publication is designed to provide accurate and
authoritative information in regard to the subject
matter covered. It is sold with the understanding that
the publisher is not engaged in rendering legal, accounting,
or other professional services. If legal advice or other
expert assistance is required, the services of a competent
professional person should be sought. *From a Declaration
of Principles jointly adopted by a Committee of the
American Bar Association and a Committee of Publishers.*

Library of Congress Cataloging-in-Publication Data

Archibald, Russell D.
    Managing high-technology programs and projects / by Russell D.
Archibald—2nd ed.
        p.   cm.
    Includes bibliographical references (p.   ) and index.
    ISBN 0-471-51327-X (cloth)
    1. Industrial project management.   2. High technology.   I. Title.
T56.8.A7   1992
658.4′04—dc20   91-25865

To
Marion

Preface

The first edition of this book was intended to fill a long-standing need for a comprehensive, unified, and practical description of the business of managing programs and projects, and of the related planning and control tools. The widespread, continued acceptance and use of that edition for the past 16 years, including translation into Japanese and Italian, indicates that it has filled that need.

This Second Edition maintains the basic structure of the book, and includes some important additions and updating to reflect the additional experience gained by senior, functional, and project managers over the intervening years in many industries and areas of application. All aspects of program/project management are treated: organizational and operational concepts and practices, the nuts and bolts of project planning and control, and many interpersonal and behavioral topics that are vital to effectively manage projects and project-driven organizations.

Part I, Executive Guide to Project Management, is of direct interest and application to general managers and other executives to whom project managers report, and their staffs. Project and functional managers, and other members of project teams, should also benefit from Part I, but will find Part II, Managing Specific Projects, to be particularly useful in carrying out their assigned responsibilities.

Six new chapters have been added to this Second Edition, 16 new sections have been added to the existing chapters, and many of the examples and illustrations have been replaced with reports and graphics from recent projects and using today's computer software packages. The new chapters are:

- **Chapter 5   The Project Team and Key Human Aspects of Project Management.** Highlights the increased recognition within the past few years of the importance of the people aspects of project management, including the need for and methods of achieving effective teamwork to enhance project success.
- **Chapter 6   Building Commitment in Project Teams.** Stresses the importance of getting and fulfilling commitment in managing projects, and provides concrete ideas for carrying this out.
- **Chapter 7   A Strategy for Overcoming Barriers to Effective Project Management.** Identifies key changes required when introducing project management practices, identifies barriers to these changes, and outlines a strategy for overcoming the barriers.
- **Chapter 11   Project Team Planning and Project Start-Up.** Experience confirms the importance of getting each project off to a well-planned, fast start. This chapter describes the methods of achieving this through team planning, and presents specific ways to conduct project start-up workshops using a case example from the telecommunications industry.
- **Chapter 13   Project Interface Management.** Positions the project manager as an interface manager, and describes proven methods of identifying, documenting, scheduling, and controlling project interfaces.
- **Chapter 14   Project Management Information Systems.** Restructures and up-dates material in the first edition and provides insight into the powerful capabilities of today's mini and microcomputer-based project planning, scheduling, cost, and resource management systems.

Successful project management requires giving up some old management habits and attitudes and learning new ways to bring together many skills to achieve the project objectives on time and within budget. This in turn requires substantial management and organizational development, and indoctrination at the executive levels, as well as indoctrination and training for project managers, functional managers and specialists, and project planning and control specialists. The book is designed to be used for this purpose, and experience in its use within government and industry, and within universities at the graduate and undergraduate levels in many parts of the world, confirms the appropriateness of its design.

The seminar/workshop approach, bringing together teams of managers and contributors from one project, preferably, or from one organization, is by far the most effective way to get people to give up the old and embrace the new. Chapter 11, Project Team Planning and Project Start-Up, presents specific, proven methods for project team training. Such seminars should use current, actual projects as case studies for application of the concepts and principles described in the book. By using real projects as the training vehicles, the seminar results are carried over into the current business of the organization. This book, augmented by appropriate case projects and case studies available elsewhere, is well suited as a textbook for university and professional courses in project management.

The book is a product of my 40 plus years of experience in many assignments around the world as an engineer, functional manager, project manager, business executive, teacher, and management consultant. During this time, I have lived in 14 of the United States; plus France, Mexico, Panama, and Venezuela; and carried out assignments in these locations plus Argentina, Australia, Austria, Belgium, Brazil, Canada, Denmark, Great Britain, Greece, India, Italy, Japan, Pakistan, The Philippines, Portugal, Spain, and Sweden. With Bendix International and ITT Corporation, my assignments concerned major programs, projects, and contracts in many locations, including Program Director of a major telecommunications effort in Mexico and of a major joint venture to manufacture automotive brakes in India. Consulting assignments have been with manufacturing, construction, petroleum, consumer product, and aerospace industries, plus a number of assignments in newly developed and developing countries. Over the years, I have conducted hundreds of seminar/workshops on project management, and in the beginning of my professional career gained experience as a project manager, project engineer, and design engineer in aerospace development and petroleum refining.

I am indebted to many people for their ideas, inputs and support in preparation of both the original book and this Second Edition. Robert B. Youker has provided many valuable contributions to both editions. I am grateful to Gerard L. Rossy for his co-authorship of Chapter 6, Building Commitment in Project Teams, and to Daniel P. Ono for his major contributions to Chapter 11, Project Team Planning and Project Start-Up. I also wish to thank David I. Cleland, Ellodee A. Cloninger, Morten Fangel, Quentin W. Fleming, Robert B. Gillis, Alan B. Harpham, Colin Hastings, Steen Lichtenberg, Sandro Miscia, David L. Pells, and John Tuman for their contributions to the book and their friendship over the years. The persons acknowledged in my preface to the first edition should again be recognized here, especially Eric Jenett, whose guidance and counsel over many years I have found to be most valuable. The contributions of David L. Wilemon, Chandra K. Jha, Marvin Flaks, Robert J. Braverman, and L. H. Brockman are again gratefully acknowledged. Discussions with many fellow members of the Project Management Institute and the INTERNET International Project Management Association, and feedback from the hundreds of participants and many clients in projects and seminars that I have been involved in have been important factors in further improvement in this book; I thank all of these who questioned, demanded, argued, and shared their valuable experiences. Once again, I should like to express my gratitude to my wife Marion for her continuing, never-failing love, interest, and support.

RUSSELL D. ARCHIBALD

Los Angeles, California
January 1992

Contents

**PART I EXECUTIVE GUIDE TO PROGRAM AND
PROJECT MANAGEMENT**

1 Project Management in Industry and Government   **3**

    1.1   Importance of Effective Project Management   4
    1.2   The Triad of Project Management Concepts   4
    1.3   Projects: Vehicles for Strategic Growth   6
    1.4   The Forces Behind the Project Management
           Approach   14
    1.5   Advantages and Costs   15
    1.6   Improving Project Management Capabilities   16
    1.7   The Organizational Impact   21

2 Programs and Projects   **24**

    2.1   Programs, Projects, and Tasks   24
    2.2   What Projects Are   25
    2.3   Comparison of Projects with Departments and
           Operations   31
    2.4   Categories of Projects   33
    2.5   Commercial and Government Projects Under
           Contract   34
    2.6   Research, Product Development, and Engineering
           Projects   35
    2.7   Construction and Other Capital Projects   35

2.8   Information System Projects   36
2.9   Management Projects   37
2.10   Multiple Projects   38
2.11   The Project Environment   38

3 Organizing the Project Management Function   43

3.1   Organizational Alternatives for Project
       Management   44
3.2   Reporting Relationships of Project Managers   47
3.3   The Manager of Project Management   48
3.4   Staffing Projects: The Project Team   49
3.5   Project Support Services   53
3.6   Charting Organizational Relationships and
       Responsibilities   60

4 The Integrative Roles in Project Management   72

4.1   The Key Integrative Roles   72
4.2   General Manager   73
4.3   Project Sponsor   73
4.4   Manager of Project Management   74
4.5   Project and Multiproject Manager   74
4.6   Functional Project Leaders   75
4.7   Responsibilities and Authority   76
4.8   Alternate Ways of Filling the Project Manager
       Role   81
4.9   Characteristics, Sources, and Selection of Project
       Managers   85
4.10   Career Development in Project Management   89

5 The Project Team and Key Human Aspects of Project
Management   91

5.1   The Project Team Concept   92
5.2   Effective Teamworking   92
5.3   Conflicts and Their Resolution   99
5.4   A Framework for Project Team Development   108

6 Building Commitment in Project Teams   109

6.1   The Importance of Commitment in the Matrix   109
6.2   Leadership and Commitment   110
6.3   Understanding Commitment   111
6.4   Key Behaviors for Managing Commitment   112
6.5   Where to Apply Commitment Building
       Behaviors   120
6.6   Balancing Commitments   122

6.7   Implementation   124
6.8   Summary and Conclusions   126

7 A Strategy for Overcoming Barriers to Effective Project Management   **127**

7.1   Key Changes Required   128
7.2   Identifying Barriers to These Changes   129
7.3   Forces Helping to Overcome the Barriers   132
7.4   Education and Training   133
7.5   Taking Appropriate Actions to Implement the Change   133
7.6   Modifying and Evolving Project Management Practices   134
7.7   Summary   134

8 Multiproject Management   **136**

8.1   Objectives   137
8.2   Multiple Major Projects Versus Multiple Small Projects   137
8.3   Project Priorities   139
8.4   Resource Management   142
8.5   Multiproject Operations Planning and Control   144

PART II MANAGING SPECIFIC PROJECTS

9 Organizing the Project Office and Project Team   **153**

9.1   Functions of the Project Office and Project Team   154
9.2   Project Manager Duties   159
9.3   Functional Project Leader Duties   163
9.4   Project Engineer Duties   164
9.5   Contract Administrator Duties   167
9.6   Project Controller Duties   171
9.7   Project Accountant Duties   174
9.8   Manufacturing Coordinator Duties   175
9.9   Field Project Manager Duties   176

10 Planning Projects   **178**

10.1   The Project Manager's Planning and Control Responsibilities   178
10.2   Project Planning and the Project Life Cycle   179
10.3   Project Objectives and Scope   181
10.4   A Project Is a Process   183

10.5   The Project Summary Plan   184
10.6   Planning and Control Functions and Tools   185
10.7   Planning During the Conceptual, Proposal or
       Pre-Investment Phases   186
10.8   Defining the Project and Its Specific Tasks:
       The Project/Work Breakdown Structure   193
10.9   Definition of Tasks (Work Control Packages)   203
10.10 Task/Responsibility Matrix   208
10.11 Interface and Milestone Event Identification   211
10.12 The Project Master Schedule and the Schedule
       Hierarchy   212
10.13 The PERT/CPM/PDM Project Level Network
       Plan   215
10.14 The Project Budget and Resource Plans   221
10.15 Task Plans, Schedules, and Budgets   225
10.16 Integrated, Detailed Task Level PERT/CPM/PDM
       Project Network Plan and Schedule   231
10.17 The Project File   235
10.18 Summary of Project Planning Steps   235

11 Project Team Planning and Project Start-Up   236

11.1   The Need for Project Team Planning   237
11.2   The Project Team Planning Process   238
11.3   Project Start-Up Workshops in the
       Telecommunications Industry: A Case Study   245
11.4   Benefits and Limitations of Project Team
       Planning   255

12 Controlling the Work, Schedule, and Costs   257

12.1   Work Authorization and Control   258
12.2   The Baseline Plan, Schedule, and Budget   266
12.3   Controlling Changes and Project Scope   268
12.4   Schedule Control   271
12.5   Cost Control   274
12.6   Integrated Schedule and Cost Control: The Earned
       Value Concept   277
12.7   C/SCSC Cost/Schedule Performance Reports   280
12.8   Technical Performance Measurement   283

13 Project Interface Management   288

13.1   Why Project Interface Management   288
13.2   The Concept: The Project Manager as the Project
       Interface Manager   289

13.3 Project Interface Management in Action 290

13.4 Product and Project Interfaces 291

13.5 Project Interface Events 293

13.6 The Five Steps of Project Interface Management 294

13.7 Conclusion 298

14 Project Management Information Systems   **299**

14.1 Defining a Project Management Information System (PMIS) 299

14.2 Computer-Supported Project Management Information Systems 302

14.3 Selection of Project Management Software Packages 310

14.4 Important Factors in PMIS Usage 316

15 Evaluating and Directing the Project   **320**

15.1 Integrated Project Evaluation: Need and Objectives 320

15.2 Methods and Practices of Project Evaluation 321

15.3 Design Reviews and Product Planning Reviews 332

15.4 Project Direction 332

15.5 Reporting to Management and the Customer 334

16 Project Close-Out or Extension   **339**

16.1 Close-Out Plan and Schedule 340

16.2 Close-Out Checklists 340

16.3 Responsibilities During Close-Out Phase 340

16.4 Project Extensions 341

16.5 Post-Completion Evaluation or Audit 342

APPENDIXES

A   **Project Start-Up Checklists**   **345**

B   **Project Close-Out Checklists**   **364**

Bibliography 371

Endnotes 373

Index 379

PART

I

‹ ›

Executive Guide to Program and Project Management

Part I is intended for top-level managers who have responsibilities for programs and projects. It should also be useful as an introduction to managers at the project level, to whom Part II is primarily addressed. Many endeavors are referred to as *programs*. This book deals with programs that are made of up *projects*. These terms are further defined in Chapter 2.

1

‹ ›

Project Management in Industry and Government

Programs and projects are of great importance to many industrial and governmental organizations. They are the means by which many companies, especially in delivering to their customers complex, advanced-technology products or systems, earn a major share of their profit. Projects are also the means by which new products are conceived, developed, and brought to market. New or improved capital facilities, as well as new information systems, are acquired through projects. Broadscope management projects, such as restructuring or reorganizing, major cost reduction efforts, plant or office relocation, and the like, are vital to continued profitable operation and growth.

In governmental units from city to county, state, regional, and federal levels, projects are vehicles for growth and improvement. School systems, universities, hospital systems, and other institutional forms of organizations create and improve their services through programs and projects. In all these various organizations—governmental, institutional and industrial—there is a growing recognition that although many projects apparently exist within the organization, they are poorly understood and seldom properly managed. The purpose of this book is to assist in the correction of this situation by presenting a concise, comprehensive and practical picture of the concepts, processes, methods, and tools of professional project management.

1.1 IMPORTANCE OF EFFECTIVE PROJECT MANAGEMENT

All projects must be well managed to achieve the desired results on schedule and within the specified cost (either in money or other critical resources). Failures in project management have caused

- Expected profit on commercial contracts to become losses through excessive costs, delays, and penalties.
- New products to be introduced late with significant detrimental impact on established business plan objectives and market penetration opportunities.
- Research and Development projects to be completed too late to benefit the related product line.
- Capital facilities to be delayed, causing missed objectives in product lines using the facilities.
- Information systems projects to exceed their planned cost and schedule, with negative impact on administration and general costs and operating efficiencies.

Failure on one significant project can eradicate the profit of a dozen well-managed projects. Too frequently the monitoring and evaluation of high exposure projects is ineffective, and the failures are not identified until it is too late to avoid undesirable results.

It is important, therefore, that every organization having responsibility for projects also has the capability to manage the projects effectively.

1.2 THE TRIAD OF PROJECT MANAGEMENT CONCEPTS

The objectives of project management are two-fold:

- To assure that the programs and projects when initially conceived and approved contain acceptable risks regarding their target objectives: technical, cost, and schedule.
- To effectively plan, control, and lead each project simultaneously with all other programs and projects so that each will achieve its approved objectives: producing the specified results on schedule and within budget.

The first of these objectives is closely linked to the strategic management of the organization, as discussed later in this chapter. Application of project management practices during the strategic planning and project concept phases has been introduced in many organizations within the past few years, with beneficial results. Too frequently, project failures can be traced directly to unrealistic technical, cost, or schedule targets, and inadequate risk analysis and management.

The project management triad contains three basic concepts that underlie professional project management:

- Identified points of integrative project responsibility.
- Integrative and predictive project planning and control systems.
- The project team: Identifying and managing the project team to integrate the efforts of all contributors to the project.

Identified Points of Integrative Responsibility

There are several places within each organization involved with projects where persons must be identified as holding integrative responsibility for a project. The most important of these are:

- The executive level: The General Manager; The Project Sponsor
- The project level: The Project Manager
- The functional department or project contributor level: The Functional Project Leaders

The *project sponsor* role is usually held by a senior manager who may or may not be the general manager. This person provides strategic direction to the project, acting for top management of the sponsoring or owner organization.[1] The *project manager* role is more operational in nature—the project manager defines the project scope and objectives to meet the project sponsor's strategic business requirements. The project manager then integrates and directs the planning and execution of the project to meet the agreed schedule, cost, and technical objectives. The several *functional project leaders* integrate the work on a given project assigned to their particular functions or subfunctions (marketing, engineering, test operations, manufacturing, production, and so on). Each of these integrative roles is discussed in more detail in later chapters, with emphasis on the project manager as the primary focal point of overall project responsibility and accountability.

Integrative and Predictive Project Planning and Control Systems

The second concept of the project management triad requires that each project be planned and controlled on an integrated basis, including all contributing functional areas or organizations, through the entire project life cycle, including all the elements of information (schedule, cost, and technical) pertinent to the situation. Most organizations are faced with the need to plan and execute many projects simultaneously using common resource pools, creating the need to use one common project planning and control system for all projects.

The Project Team

The third of the project management triad of concepts is that of designating and managing the project team, to integrate the efforts of all contributors to the project. Projects by definition consist of many diverse tasks that require the expertise and resources of a number of different specialties. These tasks are assigned to various people, usually from both within and outside the organization holding primary responsibility for the project. Other persons hold decision-making, regulatory, and approval authority over certain aspects of a project. Every individual contributing to a given project is considered a member of that project team. The most effective project management is achieved when all such contributors collaborate and work together as a well-trained team, under the integrative leadership of the project manager.

1.3 PROJECTS: VEHICLES FOR STRATEGIC GROWTH

Sources of Growth

Two basic sources of organizational growth can be identified. These are:

- *Growth by Accretion:* Slow, steady, layered growth of the basic products, services, markets, and people.
- *Step-Wise Growth:* Discrete steps—small, medium, and large—that go beyond growth by accretion.

Growth by accretion is relatively slow and most often observed in mature industries. Sales volume slowly builds, for example, perhaps as a result of the existing salesforce getting better at their jobs individually, selling the production capacity that exists within the factory, possibly assisted by more effective marketing and advertising. At some point, when the factory capacity is limiting sales (and second and third shifts have already been added) further growth is dependent on a major or step-wise change: building a new factory, or expanding the old one.

Step-wise growth occurs when the organization goes beyond growth by accretion and initiates discrete actions to expand or improve: new products or services, new markets, new processes or production facilities, new information systems, new organizational patterns, new people.

The vehicles of step-wise growth: programs and projects. Step-wise growth involves a wide range of actions from low-risk baby steps to bet-the-company giant strides. It is not possible to draw a sharp line between growth by accretion and baby steps to expand, such as hiring an additional salesperson, or taking on a new distributor in a new state for an existing product line. But when the steps become significant in size, they clearly are recognizable as projects.

Major growth steps in any organization require projects for their realization—new facilities, systems, products, services, processes, technology, markets. Acquisition of these by internal or joint ventures, purchasing another organization, merger, licensing, or other methods almost always results in a project of some complexity. More organizations are now recognizing these facts, and more are approaching the management of these growth steps using proven project management principles and practices.

The Hierarchy of Objectives > Strategies > Projects[2]

Strategically managing a company, agency, institution, or other human enterprise requires:

- A *vision* of the future of the organization at the top level;
- *Consensus and commitment* within the power structure of the organization on the mission and future direction of the organization;
- *Documentation* of the key objectives and strategies to fulfill the mission;
- *Implementation* or execution of specific programs and projects to carry out the stated strategies and reach the desired objectives.

Objectives are descriptions of where we want to go. *Strategies* are statements of how we are going to get there. Strategies are carried out and objectives are reached, when major growth steps are involved, through execution of projects and multi-project programs. Projects translate strategies into actions and objectives into realities.

It is important to recognize that objectives and strategies exist in a hierarchy—and not just at one level—in most organizations. A useful way to describe this hierarchy is to define three levels:

Level 1: Policy
Level 2: Strategic
Level 3: Operational

Figure 1.1 illustrates these levels, and shows how the strategies become objectives at the next lower level in the hierarchy, until at the operational level, projects are identified to achieve the operational objectives. Looking at the hierarchy from the bottom up, these objectives in turn represent the higher level strategies to reach the broader strategic objectives of the organization, which then support the related top-level policy objectives.

Unless the higher level objectives and strategies are translated into actions through projects, they will sit unnoticed on the shelf in the corporate planner's office, waiting for the next annual planning exercise to generate a flurry of staff wheel-spinning amid great clouds of dust and debris. It is far better for senior line managers to create their own strategies, with the implementing projects authorized through the organization's budgetary approval process; then the strategies will become realities.

Linking strategic and project management.   Figure 1.1 graphically shows the linkage between strategic and project management. Senior line managers in the organization are the creators of the growth strategies and the owners of the projects that will carry them out, with a designated project sponsor with executive responsibility for each project, and a project manager as the focal point of integrated responsibility and accountability for planning and executing each one. Strategic management sets the future course of the organization. Project management executes the specific efforts that implement the growth strategies. The managers of these projects are acting for and representing the project owners, and receive their direction through the project sponsor.

Two broad classes of organizations can be identified. In both of these types of organizations, projects are the primary vehicles for executing their growth strategies.

One, those organizations whose primary business is made up of projects. Examples of this class include architect/engineer/constructor, general contractor, and specialty contractor firms; software development firms who sell their products or services on a contract basis; telecommunications systems suppliers; and other organizations that bid for work on a project-by-project basis. Growth strategies in such organizations are reflected in the type, size, location, and nature of the projects selected for bidding, as well as the choices made in how the required resources will be provided to carry out the projects, if and when a contract is awarded or the project is otherwise approved for execution (internally, subcontract, joint venture, and so on).

The second category of organizations includes all others that provide goods and services as their mainstream business. Projects within these organizations are primarily internally sponsored and funded. Examples include manufacturing (consumer products, pharmaceuticals, engineered products, and so on), banking, transportation, communications, governmental agencies, computer hardware and software developers and suppliers, universities and other institutions, among others. These organizations depend on projects to support their primary lines of business, but projects are not their principal offering to the marketplace. Many of these sponsors of internally funded projects are important buyers of projects. These

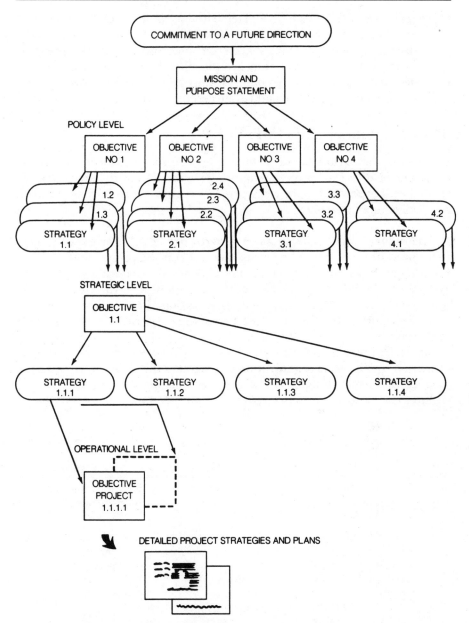

Figure 1.1 The Hierarchy of objectives strategies projects. (Source: Russell D. Archibald, "Projects: Vehicles for Strategic Growth," *Project Management Journal,* V.XIX, No. 4 (Sept. 1988), p. 32. Used by permission.)

organizations, such as the U.S. Department of Defense, the National Aeronautics and Aerospace Agency (NASA), as well as the owners in the previously mentioned categories, are program or project managers of major efforts. At the same time, they are the customers of the first category of project sellers, since they frequently must go outside their own organizations for needed expertise and resources.

The Spectrum of Growth Projects

Figure 1.2 shows representative examples of growth projects across the spectrum of different project magnitudes and serving different growth strategies. Larger and more risky projects require more formalized project management practices.

The owner's responsibilities in growth projects. For every project there must be an owner. The organization that is the prime mover behind the project is usually the owner in an impersonal, corporate sense. However, there should be a personalized owner/sponsor in the figure of one senior manager within the owner organization. In the absence of any other designation, the CEO or chairman of the corporation often falls heir to this role.

Similarly, there must be a project manager for every project. This may or may not be the same organization or person as the project owner. In the absence of formal designation of a project manager, the line manager within the sponsoring organization who manages the various contributors

Project Size	Products or Services	Markets	Profits
Small	New package Small product improvement	Add a new distributor Local advertising campaign	Substitute ingredients Increase prices
Major	Develop new product in existing line Expand existing plant Design/build new plant	Enter new markets: Domestic Foreign: —Direct —License —Joint venture	New information system Restructure organization New policies and procedures
Mega	Develop or acquire new product line	Acquire major company	Merge with competitor

Figure 1.2 Examples of projects to expand or change. (Source: Russell D. Archibald, "Projects: Vehicles for Strategic Growth," *Project Management Journal*, V.XIX, No. 4 (Sept, 1988), p. 33. Used by permission.)

to the project inherits the project manager title, whether he or she realizes it or not.

The owner responsibilities for a project are different from those of the project manager. During the planning and execution of any major project, the owner has the following responsibilities regarding management of the project.[3]

1. Provide strategic direction: Broad, executive-level decisions, oversight, and guidance:
   - Justify the need for the results of the project.
   - Develop and implement both corporate long-range strategic plans and project plans to support the project's technical performance, cost, and schedule objectives.
   - Approve the technical, cost, and schedule objectives of the project, typically summarized in the Project Summary Plan (see Chapter 10), and major changes thereto.
   - Develop and use appropriate policies and procedures to facilitate the effective management of the project.
   - Ensure use of current industry practices, knowledge, and skill in the management of the project.
   - Determine the execution strategy: Select the contributing organizations that will hold major roles in planning and executing the project, and assure that their authority and responsibilities are properly defined and understood.
   - Provide adequate information systems to facilitate the management of the project.
   - Audit performance of all parties against the plan and maintain proper surveillance of the project execution.
   - Provide decisions and approvals as required on a timely basis to assure as-planned completion.
2. Provide adequate resources:
   - Provide, on a timely basis, the money, people, and other resources required to execute the project, in accordance with the approved plan.
   - Maintain ongoing surveillance of the use of resources and replan, reprogram, and reallocate resources as necessary to keep project objectives on schedule and within budget.
3. Assure legal compliance:
   - Ensure timely compliance with all legal and regulatory requirements, including licensing and quality assurance requirements.

These basic owner responsibilities exist for any type of major growth project.

The project manager's responsibilities in growth projects.   The organization holding the project manager responsibility, and of course the individual who acts for that organization in the capacity of the Project Manager, assumes the following responsibilities:[4]

1. Achieve the project objectives:
    *   Plan, organize, staff, direct, and control the project to assure its technical, cost, and schedule objectives are achieved.
2. Organize and define responsibilities:
    *   Clearly define responsibilities and authority of all project contributors, including managers, contractors, suppliers, and consultants.
    *   Assign responsibility and authority for the quality assurance program to a senior manager who is appropriately independent from the manager holding responsibility for the technical, quality, and cost objectives of the project (when formal quality assurance practices are required).
3. Plan and control the project:
    *   Plan, schedule, and control the interfaces between all project contributors according to established systems and procedures.
    *   Establish and use appropriate project budgeting and cost control systems and procedures.
    *   Authorize and control the work, and the procurement of major equipment and materials.
    *   Develop and implement a project management information system that provides adequate information with which to manage the project.
    *   Approve (and obtain higher management approval as required) major decisions and changes in strategy, scope, schedule, and cost.
4. Comply with formal licensing and QA/QC procedures:
    *   Establish a program to assure that the licensing and quality assurance commitments agreed upon with the regulatory agencies are achieved.
    *   Assure that the quality assurance program is implemented, executed, and monitored through an organization sufficiently independent from the project execution organizations, to provide

documented assurance that the quality commitments have been achieved.

5. Evaluate and direct the project execution:

- Identify variances at the appropriate level of detail through frequent and periodic comparison of progress against plans, schedules, and budgets.
- Initiate appropriate corrective actions to assure the project objectives are achieved, or failing that, to revise the objectives.

These responsibilities and methods for achieving them are discussed throughout the remainder of the book.

Effective Project Management Produces Strategically Managed Growth

Because projects are the vehicles for step-wise strategic growth, senior managers wanting to manage their organizations strategically, that is to provide the direction and guidance for the organization to reach the long-term goals and objectives that the senior manager group has set, must provide effective project management practices linked with strategic management practices. This means that they must:

- Articulate the vision of the organization's future.
- Build a consensus within the power structure regarding that vision.
- Document the objectives and strategies to bring the vision into reality.
- Select the right projects to implement the strategies that support the higher level objectives.
- Plan and execute the selected projects effectively using proven project management practices.

In summary, organizations must continue to grow, in one or more dimensions, in order to survive over the long term. Growth can be through slow accretion or faster growth steps. Growth steps of *any magnitude* are projects. Projects having similar characteristics are often grouped into programs. Project owners have distinct responsibilities relating to their management, and these are not the same as the project manager responsibilities. Sound strategic management of the future growth of an organization requires selecting the right growth projects, fulfilling the responsibilities of the project owner, and managing them well through the use of proven project management practices.

1.4   THE FORCES BEHIND THE PROJECT MANAGEMENT APPROACH

The acceptance and use of modern formal project management concepts began in the 1950s in two diverse industries: Military/aerospace, for development of complex new systems for warfare and space exploration, and for design and construction of all types of capital facilities. Since then, project management has been discovered by many other industries. The 1980s in particular witnessed the rapid spread of the use of formal project management practices. The categories of projects identified in Chapter 2 are found in manufacturing, consumer products, banking, telecommunications, petroleum, chemical, all kinds of services, hospitals and health care, entertainment, and most other industries. As managers in these various industries recognize that they have complex projects and multi-project programs on their hands, they have adopted some or all of the project management practices described in this book.

The introduction of more formal project management practices into an established, hierarchical organization can be disruptive and can meet with considerable resistance by managers and other professionals who neither understand the need for new methods nor the project management practices themselves. In spite of this, strong forces are at work to push us toward the project management approach. The primary forces behind this phenomenon are what Galbraith calls "the imperatives of technology":

1. The span time between initiation of a project and its completion is increasing. In the past decade, major efforts have been expended to counteract this trend and shorten project durations, whether it be for developing a new product, a new system, or a new facility—with mixed results.

2. The capital committed to a project prior to actual use of the end result is increasing.

3. With increasing technology, the commitment of time and money tends to be made ever more inflexible.

4. Technology requires more and more specialized people.

5. The inevitable counterpart of specialization is more complex organization.

6. From the increased time and capital that must be committed, the increased inflexibility of this commitment, highly specialized organizations, and the problems of market performance under conditions of advanced technology, comes the necessity for more effective project planning and control.[5]

The forces pushing us toward project management emanate from the inexorable evolution of technology. These fundamental forces are not likely to diminish within the foreseeable future.

1.5 ADVANTAGES AND COSTS

The formalized project management approach presented in this book has several advantages when compared to the alternative approach of relying on the functional managers to coordinate project activities informally, using procedures and methods designed for managing their functional departments.

The basic advantages of using a formal project management approach are (1) project commitments are made only to achievable technical, cost, and schedule goals, and (2) every project is planned, scheduled, and controlled so that the commitments are achieved.

The advantages gained by appointing a project manager, or otherwise identifying a single point of integrative responsibility, include

- Placing accountability on one person for the overall results of the project.
- Assuring that decisions are made on the basis of the overall good of the project, rather than for the good of one or another contributing functional department.
- Coordinating all functional contributors to the project.
- Properly using integrated planning and control methods and the information they produce.

The advantages of *integrated planning and control* of all projects include

- Assuring that the activities of each functional area are being planned and carried out to meet the overall needs of the project.
- Assuring that the effects of favoring one project over another are known (in allocation of critical resources, for example).
- Identifying early problems that may jeopardize successful project completion, to enable effective corrective action to prevent or resolve the problems.

The advantages of effective *team-working*, especially in conjunction with the other two primary concepts of project management—focused, integrative responsibilities and integrative, predictive planning and control—include

- Bringing needed multiple disciplines together from diverse organizations to collaborate creatively to achieve project objectives.
- Bringing strong commitment and understanding to the project and its objectives.
- Developing jointly agreed plans, schedules, and budgets for executing the project, with resulting commitment to achieving the results within the target schedule and cost.
- Achieving outstanding performance on the project.

The costs related to project management include payroll and travel expenses of the project manager and supporting staff, and any costs associated with any centralized planning, control, and contract administration functions, including data processing charges.

The magnitude of the total cost of the project management function varies widely, depending on the type and size of the project and the parent organization. Experience on a variety of projects shows that the cost ranges from a fraction of one to several percent of the total project direct costs. Payroll costs of the project manager and staff are the major item.

1.6   IMPROVING PROJECT MANAGEMENT CAPABILITIES

Opportunities for improving project management capabilities can be identified and related to people, organization, processes, systems, and procedures. This section presents a general approach to identifying specific opportunities and carrying out the efforts required to capitalize on them.

Identifying Opportunities for Improvement

The need for improving project management capabilities can be determined by realistically answering these fundamental questions within a specific organization:

- Do projects (as defined in Chapter 2) exist within the organization?
- Have these projects been completed, or are they going to be completed in accordance with the original schedules, budgets, contract prices, and so on, specified in the contracts or other authorizing documents?
- Have the original profit objectives been achieved on commercial projects? Have penalties been paid?
- Can the present management structure and planning and control systems be expected to manage effectively the larger, more numerous, or

different projects required to achieve the organization's growth strategies or other long-range goals?

If the answers to these questions are affirmative, the organization's capabilities in project management should be exceptionally good. If not, various improvements are in order. These could require changes in:

- Knowledge and skills of people
- Organization of responsibilities
- Policies, processes, procedures, systems, and methods.

Recommended Approach

The material in this book is intended to assist managers in identifying and correcting ineffective practices within their organizations. The recommended approach consists of the following steps:

- Identify the symptoms of ineffective project management.
- Review the book contents to relate the symptoms to probable causes.
- Define an improvement program to attack the probable causes.
- Execute the improvement program and evaluate the results.

Project management practices are recommended to plan and execute such improvement projects.

Symptoms and Probable Causes of Poor Performance on Projects

Some *symptoms* of poor performance are:

- Late completions
- Penalties
- Cost overruns
- High project staff turnover
- Duplication of effort and inefficient use of functional specialists
- Excessive involvement of top management in project execution.

Probable *causes* of these symptoms include:

- The underlying processes (for example, the product development and launch process) are not understood or documented as integrated wholes.

- Too many projects are under way at one time for the available resources; the organization is overcommitted.
- Original schedule or cost commitments are impossible.
- No one is responsible for overall project.
- Project manager job is poorly understood.
- Project manager reports to the wrong part of the organization.
- Wrong type of person is assigned as project manager.
- Excessive conflict exists between project and functional managers.
- No integrated planning and control exists.
- There is unrealistic planning and scheduling.
- No project cost accounting ability exists.
- Project priorities are rapidly changing and conflicting.
- There is poor control over customer changes.
- There is poor control of design changes and no configuration management.
- Project offices are improperly organized and staffed.

Possible Improvement Efforts

To achieve significant improvement in an area as complex as project management, it is usually necessary to introduce changes in all areas: people, organization, processes, systems, and procedures in a well-coordinated manner.

Some typical improvement tasks in each of these areas are listed next. Additional tasks can be identified for specific situations.

1. **Management Development and Training**

   Establish development and training programs to:
   - Improve the understanding and acceptance of project management concepts and practices at all levels.
   - Develop skills required by project managers and project support specialists.
   - Create the necessary understanding of new project management policies, systems, and methods.
   - Improve the understanding and practice of teamwork.

   Develop policies and procedures related to:
   - Selection criteria for project managers, by type and size of project.
   - Career development of persons working in project management assignments.

- Performance evaluation of project managers and others assigned to or contributing to projects.

2. **Organization of Responsibilities**

- Establish a central operations planning and control office to provide integrated planning and control support for multiple small project situations.
- Establish appropriate policies regarding the roles of the project sponsor and the project manager.
- Develop responsibility matrices to clarify the relationships of all managers involved in projects.
- Develop job specifications appropriate to various types and sizes of projects, for:

    Project Manager

    Project Planner and Controller

    Project Contract Administrator

    Project Engineer

    Project Manufacturing Coordinator

    Field Project Manager

  and similar positions, as required within the organization.
- Formalize the project-functional matrix organization of responsibilities and take the actions needed to make the matrix work.

3. **Systems, Methods, and Procedures**

- Establish procedures to assure coordination of plans and actions between marketing, engineering, manufacturing and field operations: (a) prior to commitment, during submittal of a project proposal or acceptance of a contract change, and (b) during execution of the project.
- Introduce new or revised procedures to

    assure that realistic commitments are made for new projects;

    estimate and quote prices and schedules in project bids;

    authorize project work within supporting organizations and control the expenditure of project funds;

    obtain project cost accounting reports for control purposes;

    monitor and control project manpower expenditures;

    plan projects with project breakdown structures and network planning methods;

    forecast project manpower and other resource requirements;

    establish adequate project files;

carry out project evaluation and review on a systematic, disciplined basis.

- Implement integrative, multiproject information systems.
- Establish a project control room and related support procedures.

In a given situation the responsible manager should select the appropriate improvement tasks, establish their interdependencies and relative priorities, and lay out the resulting improvement program to reflect the resources available to him for the effort.

The Pilot Project Approach

The nature of project-oriented situations gives a unique opportunity to develop and test a particular group of changes on a pilot test or prototype basis, using a carefully selected project, prior to full-scale commitment to the changes. The pilot project can serve not only as a vehicle for introducing and testing new practices and methods, but also as a case study for use in management development and training efforts.

If this approach is used, care must be exercised in choosing a project that is:

- Not too far along in its life cycle.
- Representative of most other projects.
- Not so beset with inherent problems (already committed to unattainable schedules, for instance) that the benefits of my improvements cannot save it.

There is always the danger that the pilot project will receive such special attention by all concerned and therefore be so successful that the usefulness of the changes being tested cannot be determined. In this case, another result may be that other projects suffer significantly because all resources and attention have been devoted to the pilot project.

A number of improvements cannot, however, be introduced on a single project but must affect all active projects if maximum benefits are to be obtained. An example of this would be implementation of a computer-based planning and control system for multiple projects, as described in Chapter 8.

Inventory of Projects: The Project Register

An important recommended first step in any improvement effort is to prepare an inventory of programs and projects that are either in progress or in the planning or conceptual stages. Such an inventory can take the form of a *project register,* which should identify for each program or project:

- Serial number or other identifying code.
- Name of program/project manager, and percent of their time devoted to this effort.
- Customer (client, sponsor).
- Value, dollars or other currency (contract value, investment cost or other monetary measure of size).
- Key human resource investment (work-months, work-years).
- Possible exposure, dollars or other currency (penalties, loss of market, competitive gain, etc.).
- Key start and finish dates (contract award and completion, or equivalent).
- Associated projects (facilities construction, research and development, product development, other contracts).
- Date of top management review and approval to proceed.
- Other pertinent information.

A project register of this type listing all authorized projects, as well as all projects underway or planned for which formal authorization has somehow been overlooked, will provide a direct indication of whether or not the organization is simply overloaded, considering currently available resources. When too many projects have been started without careful resource planning, all projects will probably suffer delays. Without formalized project management policies and planning procedures, an overload condition can occur without the awareness of higher level management. It will probably be necessary to do some digging in the functional departments to find all the projects.

The basic responsibility of top management in this regard is to:

- Set criteria defining categories and sizes of projects.
- Require establishment and maintenance of a project register.
- Establish and revise priorities among programs and projects as required.

1.7   THE ORGANIZATIONAL IMPACT

What impact do projects have on an organization, and what is the impact of formalized project management as presented in this book?

If Galbraith is right about the imperatives of technology, then projects will have more and more impact on their parent organizations. Longer project duration, less flexible commitment of larger amounts of capital with attendant greater risks, greater specialization with its related organizational complexities—all these factors will place greater stress on our ability to manage projects effectively. And they will show up, ever more uncompromisingly, the deficiencies in present organizational structures and systems.

The internal stresses created by these continuing pressures must be reduced by relatively small changes made over a period of time—a form of stress relief that might be compared to metallurgical processes (heat treatment to prevent failure at low pressure) or geological processes (small tremors relieving stress at fault lines, preventing major earthquakes).

The grouping and regrouping of project-assigned people, equipment, and facilities—both during the life cycle of one project and when moving from project to project—does in fact enable changes to be made in responsibilities, reporting relationships, policies, and procedures on an evolutionary basis. The permanent organizational identity which is the parent of the projects can thus adjust itself to changing products and markets, while maintaining desirable stability and functional efficiency. Alvin Toffler says:

> We are witnessing not the triumph, but the breakdown of bureaucracy. We are, in fact, witnessing the arrival of a new organizational system that will increasingly challenge, and ultimately supplant, bureaucracy. This is the organization of the future. I call it "Ad-Hocracy."[6]

That is a beautiful and descriptive name for the organizational structure combining functional and project orientations in which many of us now find ourselves working. Toffler goes on to say:

> The high rate of turnover (in organizational relationships) is most dramatically symbolized by the rapid rise of what executives call "project" or "task-force" management . . .[7]

> Indeed, project management has, itself, become recognized as a specialized executive art, and there is a small but growing band of managers, both in the United States and Europe, who move from project to project, company to company, never settling down to routine or long-term operations.[8]

The first edition of this book, published in 1976, contained this statement:

> As projects become more numerous and relatively more important to their parent organizations, I believe we will see the project management function become a common feature of most large organizations.

In fact, this is what has occurred during the intervening years, with a wide variety of organizations now having well-established managers of project management; vice presidents—program management; directors of projects; and similar titles.

We can envision two types of managers in the future: those who rather permanently manage various plants or facilities, and those who are more mobile in responsibilities and manage the acquisition and execution of projects to create and produce the products to be made in the various plants. This second manager will evolve from our present day project managers and, in fact, there are some organizations that already operate in this fashion.

The rigid functional bureaucracies of yesterday and today will give way to new flexibility, but they will retain a recognizable, functional structure for the foreseeable future. We will see more and more emphasis on "temporary" management systems and structures, reflecting the needs of our temporary projects or task forces. Projects create the need for change, and the project management approach provides one mechanism for the organizational adaptation needed for the survival and growth of complex human enterprises.

2

‹ ›

Programs and Projects

Project management refers to the planning and execution of particular efforts called *projects*. The concepts and systems employed to manage projects—as well as the related difficulties—flow from the nature of the projects themselves. Therefore it is important that managers and specialists involved with programs and projects have a good understanding of their unique characteristics.

2.1 PROGRAMS, PROJECTS, AND TASKS

Some semantic confusion exists regarding these terms, which are sometimes used ambiguously and even interchangeably. Generally accepted practice in a number of industries has established the following common usage:

Program—A long-term undertaking which is usually made up of more than one project. Sometimes used synonymously with "project."

Project—A complex effort, usually less than three years in duration, made up of interrelated tasks, performed by various organizations, with a well-defined objective, schedule, and budget.

Task—A short-term effort (a few weeks to a few months) performed by one organization, which may combine with other tasks to form a project.

As a program or project is broken down into its constituent elements, there is a tendency for each organization to apply the name "project" to its portion of the overall effort, although this can cause confusion, much as the terms "system," "subsystem," and so on do in the technical area. There is probably no viable alternative in most situations, other than carefully establishing the project name in each case and emphasizing the interrelationships that exist.

The project management approach, as presented in this book, is applicable to programs, projects, or tasks. However, the primary concern is at the program and project levels. Although the term project is used throughout, application of the concepts to programs will be apparent.

Understanding of these key words is the first hurdle to be crossed in achieving good program or project management. For example, the British word *programme* frequently refers to a plan, and the Spanish *proyecto*, the French *projet*, and the Italian *progette* are frequently used to identify technical designs or plans, rather than projects as defined here. In some organizations, the word *program* can refer to a continuous effort (such as a long-term training program). This is a different use of the word than is intended in this book.

2.2   WHAT PROJECTS ARE

Examination and dissection of many projects reveal certain fundamental characteristics common to all. The important characteristics from the management viewpoint are summarized next.

Projects Are Complex Efforts That Start and End, and Are Unique

Projects are intended to produce certain specified results at a particular point in time and within an established budget. They cut across organizational lines. They are unique endeavors, not completely repetitious of any previous effort.

A Project Is the Process of Creating Specific Results

A project may be viewed as the entire *process* required to produce a new product, new plant, new system, or other specified results. The product to be created often receives more attention than the process by which it is created, but both the product and the process—the project—require effective management. The end result is not the project.

A Project Has a Life Cycle

The project life cycle has identifiable start and end points, which can be associated with a time scale. A project passes through several distinct

Type of Project \ Project Phase	1 Concept	2 Definition	3 Design	4 Development/ Manufacture	5 Application/ Installation	6 Post Completion
Telecommunication equipment contract	Pre-proposal: identification of opportunity; decision to bid	Prepare proposal; submit; receive award	Engineering design	Procure materials; fabricate, assemble, install and test	Concentration and acceptance testing	Creation of new projects for follow-up on contracts; spares; field support; final evaluation report
New product or service development project	Identify opportunity or need; establish basic feasibility	Prepare new product proposal, product plan, review & approval sheet, R&D case, project appropriation request	Design product or service; build and test prototype	Design production article; build and test tooling; produce initial production articles	Distribute and sell product; verify performance	Creation of new projects to further improve product; final evaluation report to compare results with product plan, project appropriation request, etc.
R&D case project for manufacturing development	Identify opportunity or need, establish basic feasibility	Prepare R&D case	Conduct studies, analysis & design work	Conduct pilot tests, analyze and document results	Conduct full-scale tests, analyze and document results	Final evaluation report

26

Capital facilities project	Identification of opportunity or need, establish basic feasibility	Investment analysis, budget preparation, preparation of project appropriation request	Process design, engineering for construction, equipment design and/or specification	Procure equipment, construct civil works, install and check out equipment	Start-up and commission operating facility	Final evaluation report to compare results with project appropriation request
Systems project	Identification of opportunity or need, establish basic feasibility	Investment analysis, budget preparation, preparation of project appropriation request	Systems analysis and detailed design	System coding, compiling, testing and documentation	Install and test system under production conditions	Final evaluation report to compare results with project appropriation request

Figure 2.1   Life cycle phases of various types of projects.

27

phases as it matures, as illustrated in Figure 2.1. The life cycle includes all phases from point of inception to final termination of the project. The interfaces between phases are rarely clearly separated, except in cases where proposal acceptance or formal authorization to proceed separates two phases.

The Project Character Changes in Each Life-Cycle Phase

In each succeeding phase of a project, new and different intermediate products (results) are created, with the product of one phase forming a major input to the next phase. Figure 2.2 illustrates the overall process.

The rate of expenditure of resources changes, usually increasing with succeeding phases until there is a rapid decrease at completion. The people, skills, organizations, and other resources involved in the project change in each life cycle phase. Overlapping of phases, called "fast-tracking" in engineering/construction projects and "concurrency" in military/ aerospace projects, frequently occurs. This complicates the planning and coordination needs, and places more importance on the project manager role. Key decision points occur at the completion of each project phase. Major review of the entire project occurs at the end of each phase, resulting

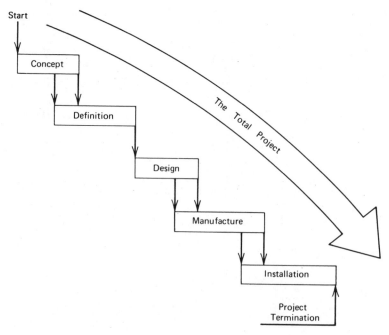

Figure 2.2   Project life-cycle phases.

in authorization to proceed with the next phase, cancellation of the project, or repetition of a previous phase.

The Uncertainty Related to Time and Cost of Completion Diminishes as the Project Matures

The specified result and the time and cost to achieve it are inseparable. The uncertainty related to each factor is reduced with completion of each succeeding life-cycle phase. Figure 2.3 illustrates the reduction of uncertainty in the ultimate time and cost for a project at the end of each phase.

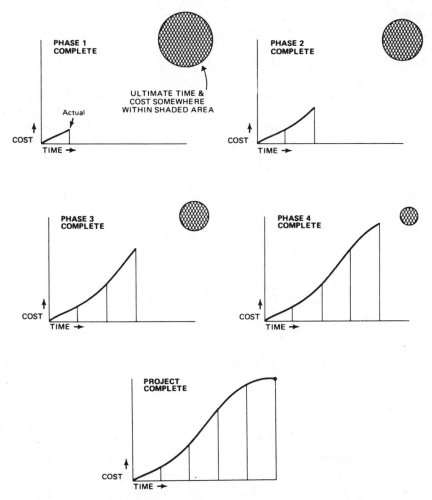

Figure 2.3   Relative uncertainty of ultimate time and cost by life-cycle phases.

In phase 1, the uncertainty is illustrated by the largest circle. The area of uncertainty is reduced with each succeeding phase until the actual point of completion is reached. The requirement for project planning and control systems and methods capable of predicting the final end point as early and as accurately as possible comes directly from this general characteristic of projects.

Many Projects Do Not Survive the Concept or Definition Phases

In almost all categories of projects in various industries, many more ideas enter the conceptual phase than emerge as fully executed projects. New product projects are canceled as they fail to meet screening criteria; proposals to design and construct new facilities are rejected, or given to a competitor. The seeds for success or failure are frequently sown in these early phases, but the outcome is not known until the project nears completion.

It is usually at the end of the definition phase, or its equivalent, when approval to spend significant amounts of money is given, and the project is reasonably assured of passing through all remaining phases. This is when the proposal is approved and a contract is awarded, or the product development funds are released, for example. It is at this time that a full-time project manager, if justified, is most often appointed, as discussed later.

The Cost of Accelerating a Project Rises Exponentially as Its Completion Nears

Recovery of lost time usually becomes increasingly more expensive for each succeeding project phase. Figure 2.4 illustrates this point for a specific type of large project. This characteristic places emphasis on the need for integrated control through all phases, with particular attention to project start-up and the earlier phases to avoid delay and shorten schedules.

The Management Implications

From these characteristics of projects, certain implications may be drawn that are important to managers:

- Projects must be managed on a life-cycle basis, with maximum continuity of responsibility and integrated planning and control from start to finish.
- Equal attention must be paid to managing both the *product* of a project and the overall process of creating that product—the *project* itself.
- Creation of a separate, self-supporting organization for each project generally is not feasible or practical, as discussed later, because of the rapidly changing situation from phase to phase.

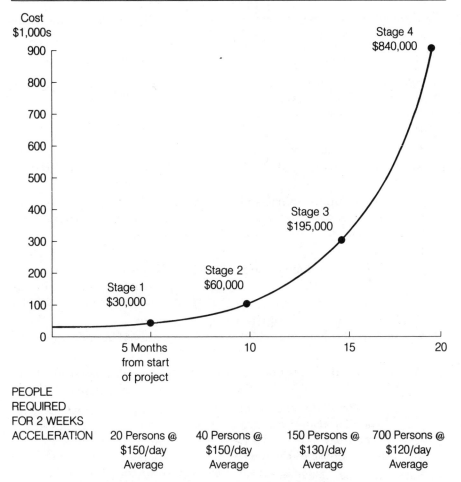

Figure 2.4   Cost of two weeks' acceleration at various project stages.

- Decisions made in early phases of a project have a greater leverage on its ultimate time and cost than those made in later phases.

2.3   COMPARISON OF PROJECTS WITH DEPARTMENTS AND OPERATIONS

A project, as identified here, is a complex, unique effort that cuts across organizational lines, has a definite start and finish point, and has specific schedule, cost, and technical objectives. A project, therefore, has important management differences when compared to a typical functional department handling repetitive work on essentially a never-ending basis, as shown in the following comparison.

Project	Accounting Department
1. Specific life cycle: conception; design; fabrication, assembly or construction; test; initial utilization.	1. Continuous life from year to year.
2. Definite start and completion points, with calendar dates.	2. No specific characteristics tied to calendar dates, other than fiscal year budgets.
3. Subject to abrupt termination if goals cannot be achieved; always terminated when project is complete.	3. Continued existence of the function usually assured, even in major reorganization.
4. Often unique, not done before.	4. Usually performing well-known function and tasks only slightly different from previous efforts.
5. Total effort must be completed within fixed budget and schedule.	5. Maximum work is performed within annual budget ceiling.
6. Prediction of ultimate time and cost is difficult.	6. Prediction of annual expenditures relatively simple.
7. Involves many skills and disciplines located in many organizations which may change from one life-cycle phase to the next.	7. Involves one or a few closely related skills and disciplines within one well-defined and stable organization.
8. Rate and type of expenditures constantly changing.	8. Relatively constant rate and type of expenditure.
9. Basically dynamic in nature.	9. Basically steady-state in nature.

Figure 2.5 illustrates the constantly changing rate of expenditure and dynamic nature of a project compared with the relatively constant rate of expenditure and steady-state nature of a purely functional department. Gilbreath states that

> Uniqueness of effort and result are the hallmarks of project situations. Consistency and uniformity are typical of operations. . . . Operations are geared to maintain and exploit, while projects are conceived to create and make exploitations available. Projects, therefore, typically precede operations in the normal business cycle. . . . If successful operation can be imagined as a continuous, uninterrupted *stream* of effort yielding a predictable collection of similar results, we must view each project as a temporary *pulse* of activity

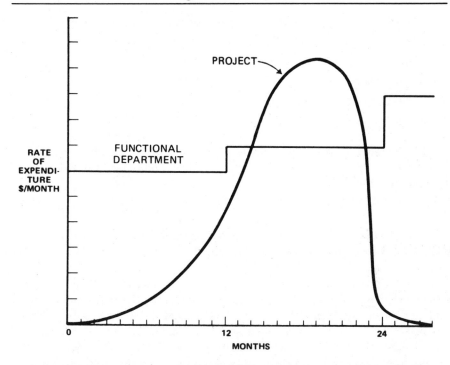

Figure 2.5   Comparison of rate of expenditure for projects and departments.

yielding a unique, singular result. . . . Operations may outlive their results, but projects expire when their result is achieved.[1]

Management Implications

Some traditional functionally oriented management concepts may not be effective when applied to projects. Methods and systems developed for planning and control of resource expenditures within functional departments are usually ineffective when applied to projects.

The relatively new concepts and systems developed to meet project needs do not in themselves create conflicts and problems, but rather they reveal the existing differences, conflicts, and incompatibilities between projects and functional organizations. A major challenge to top managers is presented by this situation: how to manage all projects effectively while maintaining the efficient operation of the functional organization.

2.4   CATEGORIES OF PROJECTS

A wide variety of projects exists within commercial and governmental organizations. These may be classified broadly into the following categories:

- Commercial projects under contract for products or services.
- Research, product development, and engineering.
- Capital facilities design and construction.
- Information systems.
- Management projects.
- Major maintenance projects (process, utilities and other industries).

The existence of projects within specific organizations varies with the nature of the business of each. Each of these categories is defined and discussed in the following sections.

2.5 COMMERCIAL AND GOVERNMENT PROJECTS UNDER CONTRACT

Projects in this category include any product or service contract having the characteristics previously ascribed to projects. This description excludes contracts of an open-ended or level-of-effort nature; contracts to deliver off-the-shelf or standard products; and contracts of a routine service nature. Commercial and government projects could involve new product development, research and development, capital facilities, and systems efforts, if performed under contract, or the equivalent within governmental documents and procedures.

Definition of Contractual Projects

A contractual project is defined as any contractual effort that involves application engineering or other customization of products or services, and that is undertaken to meet either a specific customer or internal requirement. The term is applied to customized systems designed to meet a specific need or application using existing equipment (modified to meet the specific application) or a combination of existing and new equipment. It does not refer to "off-the-shelf" stock production items sold to a standard configuration.

Identification of Major Contractual Projects

Projects in this category may vary from very small to very large, in terms of complexity, monetary value, and duration. Guidelines to aid in early identification of major commercial projects justifying formalized project management and the appointment of a full-time project manager include:

1. **Complexity**

   High degree of technical, market, manufacturing, or other complexity.

2. **Product Innovation**

   High degree of innovation or technical risk.

3. **Contributing Organizations**

   More than one separate organization contributing significantly to the project.

4. **Economic Risk**

   - 10% or more of organization's net income in any one year dependent on the project.

   - Significant penalty exposure.

   - Unusual performance guarantees.

2.6   RESEARCH, PRODUCT DEVELOPMENT, AND ENGINEERING PROJECTS

New product development includes any program or project undertaken to develop a totally new product as well as projects initiated to make major modifications to existing products.

This category includes projects to carry out well-defined research efforts (which are not simply ongoing research tasks), and projects to develop and/or engineer improved products or services, or new or improved processes, materials, or methods.

Identification of Major Research, Product Development and Engineering Projects

Criteria for identification of these major projects are the same as those listed above for contractual projects. In large developmental projects, either or both research and development and new capital facilities subprojects could be involved. In this case the overall project would be considered a new product development project, with appropriate regard for the different characteristics of the research and development and capital facilities portions, as discussed in following sections.

2.7   CONSTRUCTION AND OTHER CAPITAL PROJECTS

Most organizations have well-established policies and procedures for controlling projects requiring investment of capital funds. These policies and procedures provide detailed definitions of various types of capital projects. Generally, these involve expenditures to acquire land, buildings, and capital equipment, whether by purchase, construction, or lease; and extraordinary expenditures for major alterations and rearrangements of existing facilities.

When a capital project is indispensable to or forms a part of a customer contract, new product development, research, development, and engineering, information system, or management project, it should be managed as an element of the total program or project. Usually, additional financial controls are required for capital projects, to assure that the proper costs are capitalized as desired by management and required by accepted accounting practices.

Identification of Major Capital Projects

Criteria for major capital facilities construction projects usually requiring a full-time project manager and a formal project management approach are:

1. **Complexity**

   High degree of technical or operational complexity, such as interaction with ongoing production operations.

2. **Economic Risk**

   - Facilities required to support a major commercial, new product or research, development, and engineering project.

   - Investment of $3 million or more for civil works or equipment.

Capital expenditures to lease or purchase land, facilities, or equipment would normally be projects of short duration covering the negotiation period. These would usually be handled by functional specialists rather than as a major design and construction project requiring a full-time project manager.

2.8   INFORMATION SYSTEM PROJECTS

Information system projects are specific undertakings related to any segment of the organization's business resulting in the use or commitment of general systems or information processing equipment, personnel, or other resources.

General systems refers to techniques such as work simplification, work measurement, forms coordination and control communication analysis, operations research, company manuals, and procedures records retention, and so on.

Information processing refers to all types of data processing equipment, software packages, outside consultants, internal personnel, and so on involved with information processing activities.

Identification of Major System Projects

Criteria for identification of major system projects, which would usually require a full-time project manager, are:

1. **Complexity**

   High degree of technical or user complexity.

2. **Innovation**

   High degree of technical innovation and risk.

3. **Contributing Organizations**

   More than one separate organization contributing significantly to the project.

4. **Economic Risk**

   - System required to support a major commercial, new product, research, development, and engineering or capital facilities project.

   - Investment of 10 person years or more of professional effort.

2.9 MANAGEMENT PROJECTS

Many types of management projects may be encountered in various organizations. Frequently, these are assigned to an ad hoc committee. They include:

- Major cost reduction efforts.
- Reorganizations.
- Specific mergers, acquisitions, or divestitures.
- Major geographic or market expansions.

Other examples are specific to situations within particular organizations.

Management projects often become the spawning ground for other projects in the categories previously described. It is not possible to establish criteria to identify major projects in this category, because of the widely varying nature and importance of each specific effort.

2.10   MULTIPLE PROJECTS

Within many organizations, a relatively large number of identifiable but smaller projects exist. Each of these projects, taken by itself, could be managed effectively without introduction of more formalized project management concepts. But because of the number of such projects in progress at any one time, many are not effectively managed, as evidenced by delays, reduced profits, or overruns in their approved budgets.

Definition of a Multiproject Environment

When several small projects exist (in any of the categories described in the preceding sections) within an organization and these projects in the aggregate represent an economic risk equal to or greater than that represented by one major project, then it is proper to term this a multiproject situation.

It is difficult to draw a sharp line between a multiproject environment and a multicontract situation that involves routine sales, production, and delivery of products. If each contract is basically a repetition of previous ones, with little change or innovation, then the contracts should probably not be considered projects. If there is a large amount of redesign, or customizing, then it may be beneficial to identify each contract as a project.

A good example of this is the telephone switching business. The production and delivery of standard items of equipment is not of a project nature. However, the design, production, installation, testing, and delivery of a completed, functionally operating telephone exchange quite definitely *is* a project. Such a project would of course depend upon the repetitive production of many standard parts, as well as some specially designed parts in most cases.

Multiproject Management

While it is not feasible to assign a project manager for every telephone exchange delivered by a company, the basic concepts presented in this book are useful in this situation. In each chapter, appropriate distinction is made between application of the principles and system to a small number of major projects and to larger numbers of relatively small projects. Chapter 8 discusses the multiproject requirements in more detail.

2.11   THE PROJECT ENVIRONMENT

Any project must be understood and managed within the particular environment in which it exists. What works well for a specific project within one environment may be ineffective in a different environment. Many projects have difficulties and are less successful than desired because their objectives, organizational design, and management approaches are incompatible or in conflict with key environmental elements.

Project success will be determined in many situations not so much by the most logical or efficient arrangement of roles, responsibilities, and resources, but by the most workable co-alignment of the various internal parts of the project with external agencies. These could include local, regional, and national government agencies; suppliers of goods and services; users of the project results; and, most importantly, those who benefit the most (or are most affected by) project success—the project beneficiaries.

It is important for the project manager first to understand the project environment, and second to link the project with the key actors and factors in the environment in such a way that project success is enhanced to the greatest extent possible. Finally, the type and magnitude of change within the environment must be anticipated, and sufficient flexibility provided in the management approach to accommodate these changes.

The Environmental Scan

Not all elements in a project's environment will be crucial to its success. Systematic scanning of the environment to identify the key actors and factors which are crucial is an important part of the project manager's, and indeed the entire project team's, roles. Such scanning can vary from an undirected, fortuitous, and subconscious observation to a purposeful, predetermined, and highly structured inspection. Figure 2.6 presents a format that has proven to be useful for this purpose. In using this approach, a fairly large number of potentially crucial environmental actors and factors are first identified and noted in the appropriate sector and ring. Then, usually through project team consensus, the few most critical actors and factors are highlighted as shown, by outlining either with a square box or an oval. *Actors* are defined as individuals, groups, institutions, organizations, agencies, and so on who can take action (or fail to do so) affecting the project. *Factors* are elements such as attitudes, laws, regulations, customs, habits, trends, physical or economic conditions, and so on which, although they cannot take actions, nevertheless exert great influence on the project by their very existence.

Linkages Between the Project and Its Environment

To be successful, the project must be linked wherever possible to its environment. This is accomplished through linkages with the key actors and factors that have been identified in the environmental scanning. These linkages can be made through organization structures and management processes.

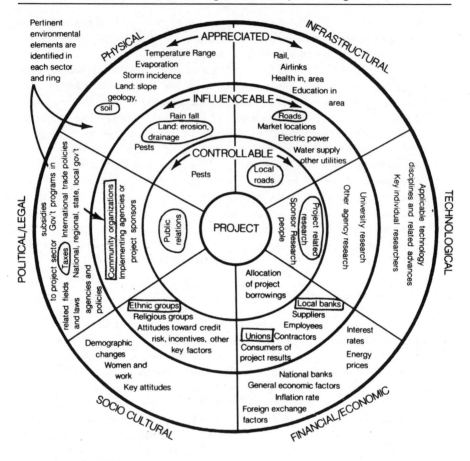

Key Actors: ▭
Key Factors: ⬭
(for a specific project)

Figure 2.6 A method for documenting an environmental scan. (Adapted from Nicholas R. Burnett and Robert Youker, "Analyzing the Project Environment," CN-848, July 1980, The Economic Development Institute (World Bank) Course Note Series, Washington, DC.)

Project Management Tasks	Key Environmental Actors and Factors	
	Actors	Factors
1. Define the project.	Get them involved in the definition process; use their ideas if possible; explain the project to them; identify problems and negative reactions.	Assure that key factors are included in planning assumptions; identify limits imposed by key factors which affect project definition.
2. Organize and build the project team.	Establish formal, ad hoc and informal relationships with key actors; consider actors as members of the project team; include where possible in meetings.	Incorporate the influence in organization design; make all team members aware of the key factors and their effects on project success.
3. Plan, schedule and budget the work.	Involve actors in preparation of plans, schedules and budgets, as possible; assure that plans reflect realities imposed by key actors.	Incorporate information relating to key factors into plans, schedules and budgets.
4. Authorize the work and start up the execution activities.	Keep actors informed, especially when activities will have a direct impact on them.	Monitor factors to avoid direct conflicts and generation of problems whenever possible
5. Control the work, schedule and costs.	As above.	As above.
6. Evaluate progress and direct the project.	Include actors in evaluation process where possible; give advance information on major changes.	Periodically update data on each key factor, and include in progress evaluation process.
7. Close out the project.	Involve actors in close-out planning and activities; keep informed as appropriate.	Continue to monitor factors and reflect changes in close-out plans.

Figure 2.7 Examples of linkages between the project and its environmental actors and factors.

Here are some examples of organizational linkage:

- Formal steering committees.
- Electing key actors to the board of directors or advisors to the project or its sponsor.
- Arranging for the project manager or project team members to serve on the boards or on committees of the key actor's organizations.
- Teaming up project people with the key actors in third-party organizations.
- Ad hoc committees for coordination or planning, with key actor participation.
- Appointing liaison managers with responsibilities to interact with the key actors, or having the project manager perform this function.

Here are some examples of management process linkages:

- *Actors:* Invite key actors to participate in project review meetings, especially when developing plans that affect them. Invite key actors to tour the project offices and field sites, and introduce them to project team members and visiting dignitaries. Provide them with copies of appropriate reports.
- *Factors:* Assure that the identified key factors are properly considered when developing project plans, budgets, schedules, and so on. Establish reporting systems so that changes can be reflected in project plans. Establish systems to monitor assumptions and alert the project manager to significant changes.

Figure 2.7 presents some further examples of how the project can be linked to its environment during seven generic project management tasks.

Dynamics of the Project Environment

Just as the project itself is changing in character as it moves through its life-cycle phases, so too is the project environment changing. Project managers must re-examine their key environmental elements periodically to re-affirm the identity of the key actors and factors which relate to their projects.

3

‹ ›

Organizing the Project Management Function

Creating the proper organizational setting for project management must be accomplished in the absence of well-known principles, such as those that apply when organizing a traditional, functionally oriented structure. No single organizational pattern has yet emerged to answer the following project management questions:

- How will the project manager and other integrative responsibilities be assigned (as discussed in Chapter 4)?
- To whom should the project manager report? At what level, and within which part of the organization?
- Who should be assigned as full-time project office members reporting only to the project manager, and who should contribute as project participants while remaining in their functional departments?
- How are specialist staff skills in project planning and control, contract administration, finance, legal, and so on, best provided to project managers?
- Who is responsible for development and operation of multiproject, integrated project planning and control systems (discussed in later chapters)?
- Who should hold specific responsibility for multiproject management?

In this chapter, the underlying factors influencing the answers to these questions are summarized, and, where possible, some basic guidelines are set forth. Illustrations are provided of various organizational arrangements used by a number of different companies. The objective of this chapter is to enable the responsible executives to create the most effective organizational arrangement for their projects within their own organizations, with proper assistance of staff specialists in project management and organizational planning, and to enable affected managers and specialists to understand and more fully accept the resulting arrangements.

3.1 ORGANIZATIONAL ALTERNATIVES FOR PROJECT MANAGEMENT

Three basic alternative forms of organization have been used for the planning and execution of projects: (1) purely functional, (2) function/project matrix, and (3) purely project (that is, a functional organization devoted solely to one project). Each of these has its strengths and weaknesses. Each can be made to work, with varying degrees of effectiveness, depending upon the characteristics of the basic organization (size, degree of rigidity, nature of the business, culture, and habits) and of the projects (number, size, complexity, degree of uniqueness, duration, and other factors previously discussed in Chapter 2).

Companies and agencies typically evolve their approach to managing their projects through some combination of these basic forms. Initially, most organizations are structured along the classic functional pyramid lines, with separate departments for marketing, engineering, financial, manufacturing, or other operations, and staff specialists for legal, treasury, human resources, administration, and so on. (Product lines, geography, technologies, and customers are often also represented in the pyramidal structure.) As projects within the functional organization become more numerous, larger, and more complex, and as schedule and/or cost performance becomes more critical, the managers introduce changes which lead them either to a functional/project matrix solution or to establish essentially stand-alone, "projectized" organizations. Often, a series of failures in the purely functional organization (cost overruns, missed schedules) forces the senior managers to look for a better way. It is rarely possible to establish a completely separate organization for each and every project, due to the cost of duplicating all the needed specialized skills and facilities. The result is that most organizations find themselves in a function/project matrix of some kind.

Youker has summed up the dilemma very succinctly:

The functional, hierarchical organization is organized around technical inputs, such as engineering and marketing. The project organization is a single purpose structure organized around project outputs, such as a new dam or a new product. Both of these are unidimensional structures in a multidimensional world. The problem in each is to get a proper balance between the long-term objective of functional departments in building technical expertise and the short-term objectives of the project.

Matrix Organizations

The matrix organization is a multidimensional structure that tries to maximize the strengths and minimize the weaknesses of both the project and functional structures. It combines the standard vertical hierarchical structure with a superimposed lateral or horizontal structure of a project coordinator [or manager].

The major benefits of the matrix organization are the balancing of objectives, the coordination across functional lines, and the visibility of the project objectives through the project coordinator's [manager's] office. The major disadvantage is that the [person] in the middle is working for two bosses. Vertically, s/he reports to the functional department head. Horizontally, he reports to the project coordinator or project manager. In a conflict situation, he can be caught in the middle.

The project manager often feels that he has little authority with regard to the functional departments. On the other hand, the functional department head often feels that the project coordinator is interfering with his territory.

The solution to this problem is to define the roles, responsibility, and authority of each of the actors clearly. The project coordinator specifies what is to be done and the functional department is responsible for how it is done.[1]

The roles, responsibilities, and authority of the various people involved in project management within the matrix organization are discussed in detail in following chapters.

The Project Taskforce

Frequently, organizations find it useful to physically locate a large part of the project team in one place, to enhance communication, control, and teamworking. Some may view this as a "projectized" organization, but many times it is still a matrix, since the functional contributors are not always transferred formally from their home departments to the project department.

Weak to Strong Matrix—A Continuum

In Youker's words,

The three major organizational forms—functional, matrix, and project— may be presented as a continuum ranging from functional at one end to project at the other end (Figure 3.1). The matrix form falls in between and includes a wide variety of structures, from a weak matrix near functional to a strong matrix near project. The continuum in Figure 3.1 is based on the percentage of personnel who work in their own functional department versus the percentage of personnel who are full-time members of the project [organization] . . . The dividing line between functional and matrix is the point at which an individual is appointed with part-time responsibility for coordinating across functional department lines.

The bottom line of Figure 3.1 shows that a weak matrix has a part-time coordinator. The matrix gets stronger as you move from full-time coordinator to full-time project manager and finally to a project office that includes such personnel as systems engineers, cost analysts, and schedule analysts. The difference between a coordinator and a manager is the difference between mere integration and actual decision making.

On the far right we have the project organization. Ordinarily, there is a clear distinction between a strong matrix in which most of the work is still being performed in the functional departments and a project organization in which the majority of the personnel are on the project team [in the project organization].

It is rare for a project organization to have all the personnel on its team. Usually some functions, such as accounting or maintenance, would still be performed by the functional structure. . . .

Strong and weak are not used in the sense of good and bad. Rather, they refer to the relative size and power of the integrative function in the matrix.[2]

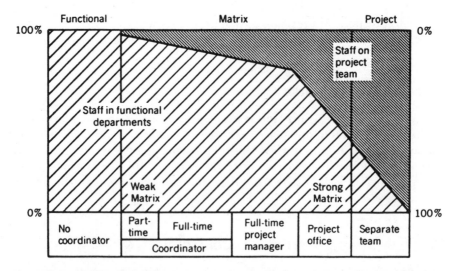

Figure 3.1   Organizational Continuum 1. (Source: Robert Youker, "Organizational Alternatives for Project Management," *Project Management Quarterly*, V. VIII, No. 1 (March 1975). Used by permission.)

3.2 REPORTING RELATIONSHIPS OF PROJECT MANAGERS

After a major project has been identified and the decision made to assign a project manager, the next decision regarding the level in the organization and the specific part of the organization in which he will be placed is crucial to his effectiveness.

Basic Principle of Reporting Location

The basic principle which seems to apply is that *the project management function should report to the line executive or manager who will actually resolve essentially all of the conflicts within and between the projects to be managed.*

Thus, a project manager could properly report to the manager of a department, division, product line, or to the general manager or managing director of a company. For very large projects involving several companies, he or she would probably report to the managing director of the "lead house," which has prime contractual responsibility.

Other Important Factors

The size, scope, and nature of the project are key factors in determining where the project manager should report. The number of projects within the organization and the existence or absence of a formalized multiproject responsibility will also influence the decision. The executive rank, seniority, experience, and personality of the project manager will also have a direct effect on the final decision regarding his reporting location.

High versus Low Reporting Locations

Project managers traditionally feel that their job would be easier if they reported to the highest executive in the organization. Experience indicates that this is not always a valid conclusion. A reporting level that is too high can be as ineffective as one that is too low, for different reasons.

If the reporting level is too high, the result can be serious and unneeded conflict with senior line managers, impeded communication with functional managers contributing to the project, or unresponsiveness to the project manager caused by the separation of several levels of organization, causing him to resort to using the power of his reporting position, further increasing conflict and retarding communication.

If too low, as for instance reporting to a department (or lower) manager when four other departments are contributing significantly to the project, the result can be inability of the project manager to get needed cooperation from the other departments, or inability to evaluate critically and objectively the contribution of the department to which he is assigned.

In either case, the project will suffer because the project manager is not located at the level commensurate with the needs of the project.

3.3   THE MANAGER OF PROJECT MANAGEMENT

When an organization has responsibility for a number of projects of relatively similar nature, there are two possible alternatives regarding the reporting location of the various project managers:

- The project managers may individually report to line executives or managers, as appropriate to each project.
- One executive or manager may be given the multiproject responsibility, and all project managers may report to him (with exceptions for very large or important projects, very senior project managers, or relatively small projects).

The second alternative may result in establishment of a "project (or program) management office," under a Manager or Director of Project (or Program) Management who reports to a senior executive. A definite trend is observed over recent years to set up such an office in a number of industries. This reflects the coming of age of the disciplines of project management, which is taking its place as an area of functional expertise, along with engineering, marketing, manufacturing, and so on.

The Manager of Project Management, or equivalent title, typically holds responsibility for:

- Providing day-to-day direction to the project managers and project support specialists assigned to the project management department or office.
- Resolving priority, resource, and other conflicts, or escalating such conflicts as necessary for resolution.
- Selecting and developing project managers and project management specialists, and establishing more stable career paths for them.
- Developing and implementing improved project management practices and information systems, and training all concerned in their use.
- Informing senior managers of progress and problems on all projects.

The project managers in such an office may have only one major project assigned to each of them or each may be responsible for a number of smaller projects, as discussed in Chapter 8.

3.4   STAFFING PROJECTS: THE PROJECT TEAM

Alternative Staffing Methods

The three basic alternatives (ordinarily used in combination) in staffing a project are:

- Assign people directly to the project office, under the control of the project manager.
- Assign tasks required for the project to specific functional depart- ments or specialized staffs.
- Contract for project tasks with outside organizations.

The project manager typically desires to have all the people contribut- ing to his project assigned full-time to the project organization or office that he directly manages. In most situations, this is undesirable, if not impossible, because:

- Skills required by the project vary considerably as the project matures through each of its life-cycle phases.
- Building up a large, permanently assigned project office for each project inevitably causes duplication of certain skills (often those in short supply), carrying of people who are not needed on a full-time basis or for a long period, and personnel difficulties in reassignment.
- The project manager may tend to be diverted from his primary task and become the project engineer, for example, as well as having to become concerned with the supervision, administration, and person- nel problems of a large office rather than concentrating on managing all aspects of the project itself.
- Professionally trained people often prefer to work within a group devoted to their professional area, with permanent management hav- ing qualifications in the same field, rather than to be isolated from their specialty peers by being assigned to a project staff.
- Projects are subject to sudden shifts in priority or even to cancella- tion, and full-time members of a project office are thus exposed to potentially serious threats to their job security; this often causes a reluctance on the part of some people to accept a project assignment.

All of these factors favor keeping the full-time project office as small as possible and depending on established functional departments and spe- cialized staffs to the greatest extent possible for performance of the vari- ous tasks necessary to completing the project. This approach places great emphasis on the planning and control procedures used on the project.

On the other hand, there are valid reasons for assigning particular persons of various specialties to the project office. These specialties usually include:

- Systems analysis and engineering (or equivalent technical discipline), and product quality and configuration control if the product requires such an effort.
- Project planning, scheduling, control, and administrative support.

Experience of many companies indicates that at least these specialties must be under the direct control of the project manager if he is to be able to carry out his assigned responsibilities effectively. Setting up a large project office may be done to achieve better control of the work, in the absence of effective planning and control procedures or systems.

Organization of Project Participants

The team of project participants includes all persons to whom specific project tasks have been assigned, including these:

- Directly under the project manager in the project office.
- In functional departments.
- On specialized staffs.
- In outside organizations.

Figure 3.2 illustrates a generalized organization for an industrial project showing all participants and indicating the persons and functions desirable to have in the project office (under the direct authority of the project manager), and those participants almost always indirectly related to the project manager. The persons and facilities that may vary in their reporting relationship, depending on the situation, are also indicated, including:

- Prototype design, fabrication, and test.
- Production design.
- Manufacturing project coordinator.
- Field project manager.

Some or all members of the product design group are justified in being assigned to the project office if

- Continuous, close communication is required with other members of the project office.
- They are needed for extended periods of time (for example, 6 months or more).

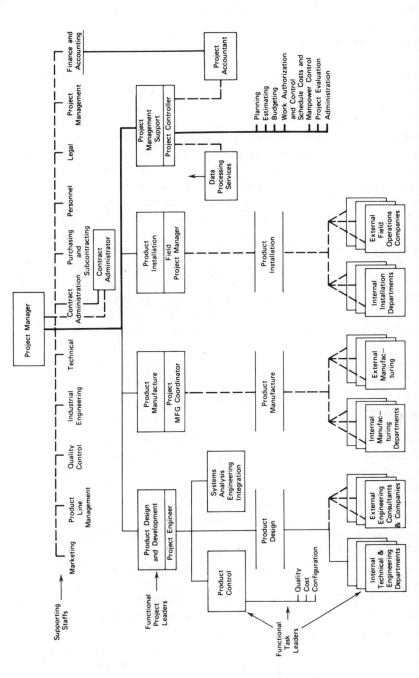

Figure 3.2 Generalized organization of the project team. ⎯⎯ Project office members recommend in most cases; ⎯ ⎯ may or may not be members of project office; ⎯⎯ not usually members of project office.

- There is otherwise a low confidence that they will be able to spend the agreed upon time and effort on the project. In other words, inadequate control exists.

Extreme care is required to select only those persons who are definitely justified in being assigned to the project office before such transfers are made.

A practical alternative is to identify those engineers and other specialists within the functional departments who are to perform the assigned project tasks, and physically group these people together, either within the functional department or within the project office. In this case, the people are not formally transferred to the project office but maintain their permanent relationships to their functional departments.

The manufacturing project coordinator should probably remain within the functional manufacturing organization because of the need for close coordination and contact with the manufacturing people.

The field project manager will probably retain a permanent relationship with the field organization, such as the installation or other field operations department, and report to the project manager on an indirect basis, especially if the project is relatively short or if the field project manager is involved with several projects at once. When this phase of the project covers an extended period (6 months or more), it is recommended that the field project manager be assigned to the project office.

Effect of Staffing Method on the Project Manager's Authority

If all required people are placed directly under the project manager, his role is quite similar to that of a multifunctional division manager. For the reasons discussed this rarely occurs, except in very large, high priority projects.

In practice today, much (and in many cases, most) of the work is performed for a project by people in various functional departments or outside the organization under contract. These people do not report organizationally to the project manager, but he must still integrate their efforts to achieve the project objectives. The result is the situation discussed in Chapter 4, where task leaders are receiving direction from both a functional boss and a project boss. The authority of the project manager in this project/function situation is changed considerably from that which he has on a fully autonomous project.

Under these conditions, effective project management requires that the project manager act as an "interface manager," and that adequate project support services be provided to the project manager to establish his control of the project through integrated planning, scheduling, and evaluation.

Relationships Between Project and Functional Managers: Interface Management

It is difficult to convey a good understanding of the project manager role to functional managers whose experience has been wholly within traditional, functional organizations. The project manager's responsibilities are integrative and relate to the interaction between the various contributing functions for his or her specific project. These responsibilities overlay those of the functional managers but do not change the basic accountability of each functional manager for his or her portion of the project.

If the general manager, to whom all the functional managers report, were to retain the project manager role himself, then the functional manager would readily understand and accept his role as project manager. But when responsibility for the project is assigned to someone other than the general manager, then the functional managers may have difficulty accepting the idea.

The concept of the project manager as an "interface manager" is useful to clarify his relationships with functional managers and others outside his direct control. Within this concept, the project manager's prime responsibilities are to:

- Identify the various interfaces between functional departments and other elements of the project.
- Develop plans and schedules that incorporate these interface points.
- Communicate the current and future status of all interfaces to all affected functional contributors.
- Monitor progress in all areas and periodically evaluate the project to identify problems and initiate appropriate corrective action.

Interface management is described more fully in Chapter 13. Project managers who practice interface management will find they do not need to invade the prerogatives of the functional managers, and their relations will noticeably improve. The project manager can, in fact, be of such help to the functional managers in this manner that his role will be welcomed enthusiastically.

3.5 PROJECT SUPPORT SERVICES

In addition to the people and facilities required to design, manufacture, and install (or otherwise put into use or operation) the product to be created, specialized project management support services are needed to assist the project manager in carrying out his or her responsibilities. As described in Chapter 4, these responsibilities include achieving the technical (product) objectives within the established limits of time and cost. The specialized support services relate to both the *product* and *project* planning and control functions described in more detail in Chapter 10. Figure 3.2 illustrates the organization of the support services.

Product Support Services

The services of greatest importance to the product, and which may be provided by assigned members of the project office, are:

- **Technical Management**

  Direct the product design and development work.

- **Systems Analysis, Engineering, and Integration**

  Analyze and specify product performance requirements and criteria.

  Evaluate and integrate detailed design by all project participants.

  Design and specify the system and product functional tests and evaluate test results.

  Carry out design review practices.

- **Product Control**

  Establish detailed performance objectives and assure achievement in product quality, reliability, and maintainability.

  Document a base-line system and product configuration, and control and document any changes to that configuration.

  Establish detailed estimates of product cost and revise these as design changes or other revised information become available.

Project Management Support Services

These include specialized functions related to the following:

- *Contract administration,* as described in detail in Chapter 9, usually performed by a contract administrator assigned to the project but not considered a member of the project office under the direct control of the project manager.

- *Project accounting,* performed by a member of the accounting department designated for the project, and usually reporting indirectly to the project controller (or equivalent).

- *Specialized project management support services,* usually performed by members of the project office under the direction of a project controller (or similar title). These include:

Planning: Project definition, work planning.

Estimating: Determining the work hours and other resource requirements for all project participants; costing the resource requirements.

Interface identification: Identifying, communicating and managing the points of interrelationships between contributing organizations and between various project elements.

Budgeting: Allocating approved amounts of work hours and money resources to all project participants.

Controlling the scope of work.

Work authorization and control.

Schedule control.

Cost control.

Project monitoring and evaluation: Services supporting the practices described in Chapter 12.

Administration: Establishing and maintaining the project file; other administrative support required by the project manager.

Specialized data processing support services, provided directly to the project office by available data processing staff and facilities, in support of project management systems.

Powerful microcomputer software packages for project planning and control enable the project manager and team to be self-sufficient in many situations. (See Chapter 13.)

Size of the Project Management Support Staff

The number of people required to carry out the management support functions will vary, depending on the size and complexity of the project, from one person (the project manager) with a secretary, to 10, 15, or more on very large projects. For example, Figure 3.3 shows the persons assigned to a major electronic development project.

In general, it is desirable to hold the project office to a minimum number of people. Further discussion of assignment of persons to the project office is given in Chapter 9.

Centralized versus Decentralized Project Management Support Services

In situations involving a number of relatively small projects, it is not possible to provide a project manager and supporting project staff for each. In this case, a centralized project management support group is required to

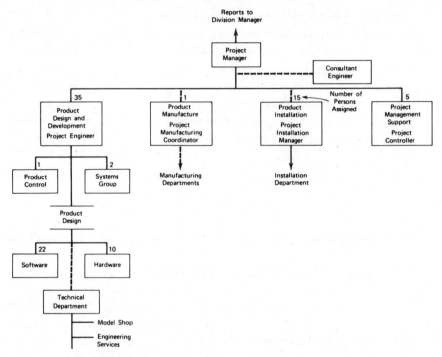

Figure 3.3   Examples of organization for a major electronic project under customer contract. ____ Members of project offices; ____ project participants not reporting directly to project manager. The project manager also has the usual staff support.

provide the planning and control services for whoever is responsible for the projects, whether this is a line manager, a project manager, or a multi-project manager.

Experience indicates that a centralized group is less effective in support of those larger projects that could justify one or more full-time people performing support functions. Because these functions deal with vital information regarding the current health and ultimate success of the project, a project manager on a large effort will not entrust them to an outside group. If these services are offered or imposed from the outside, the project manager will generally shield the outsiders from key information and ignore their efforts of assistance. This often results in developing internal methods for performing these functions, which may be less effective and more costly than if qualified specialists were placed within the project office.

Central Planning for Multiprojects

In most large companies, major projects exist together with numerous smaller projects. Thus, there is frequently a need for a centralized project support staff to plan and control the smaller projects, as well as for equivalent specialists assigned to the various major project offices. The central staff can be used as a training ground for persons to be assigned to major project offices and to coordinate development of improved methods, procedures, and systems, including multiproject planning, resource allocation, and control systems.

The concept of a full-blown operations planning and control function, supported by a computer-based operations planning and control system, is presented in Chapter 8. Where such a function exists, it can provide very effective planning and control support to the various project managers.

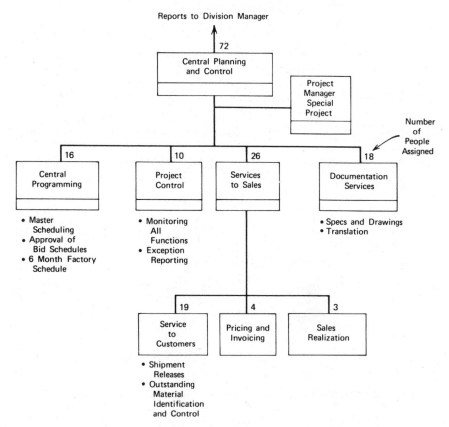

Figure 3.4   Example of central planning and control for a major division.

Figure 3.4 shows an example of this type of multiproject management support in a large telecommunications company.

Relationships Between Project Office, Central Planning and Control, and Information Processing Services

Effective use of advanced systems such as those described in Chapter 14 requires skilled information processing support services. These services must be made available as directly as possible to the project managers or

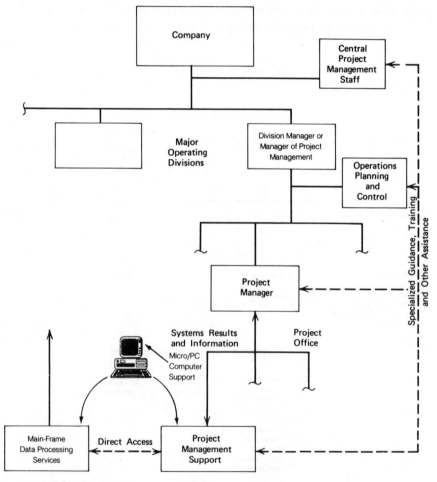

Figure 3.5   Recommended relationships between central staff, opertations planning and control, project office, and data processing services.

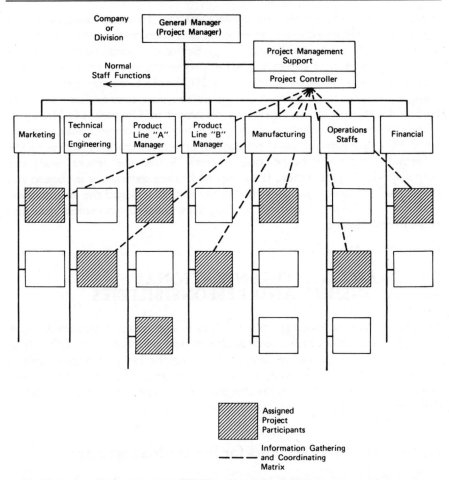

Figure 3.6   Generalized organization chart with project manager responsibility retained by line manager. The matrix is limited to fact finding, assisting in the flow of information, and coordinating to execute decisions communicated through the normal reporting channels by the project manager/line manager.

project controllers within the project offices. In some companies, the central staff responsible for developing and encouraging the use of such systems has insisted on acting as the go-between, funneling the information from projects to the information processing department and back. The availability of user-friendly, powerful project management systems using mini- and microcomputers within the past decade has overcome this problem to a great extent, as discussed in Chapter 14. However, where large mainframe computers are required, project management support people should insist on having direct, interactive access to the information and the supporting systems. While the central staff can provide valuable guidance, training, and other assistance it should not insert itself between the users in the project offices and the information processing services. Figure 3.5 illustrates the recommended relationships in this regard.

3.6 CHARTING ORGANIZATIONAL RELATIONSHIPS AND RESPONSIBILITIES

Because the project manager responsibilities and relationships do not conform to traditional organizational theory and practice, it is difficult to illustrate these relationships accurately using the familiar organizational charting methods. Various organizations have prepared different kinds of charts in attempting to portray the project management situation graphically, to enable analysis and improved understanding.

Alternative Arrangements for Project Management

Various alternative organizational arrangements have been and continue to be used for the project management function. Most of these result in what is usually termed a "matrix" organization, wherein the project manager's relationships are overlaid on the basic organizational structure, forming a matrix of reporting, direction giving, and coordinating lines of communication and linkage.

The most common forms of the resulting organization, as illustrated with the usual charting methods, are shown in Figures 3.6 through 3.15.

Responsibility Matrix

Traditional organization charts and position descriptions are necessary and valuable, but they do not show how the organization really works.

Another approach which comes closer to this goal has evolved and is generally referred to as *linear responsibility charting* to produce a responsibility matrix. To be most effective, the members of a work group should actively participate in developing such a chart to describe their roles and relationships. Such development resolves differences and improves communications so that the organization works more effectively. The responsibility matrix is useful for analyzing and portraying any organization, but it is particularly effective in relating project responsibilities to the existing organization. This important tool is described in more detail in Chapter 10.

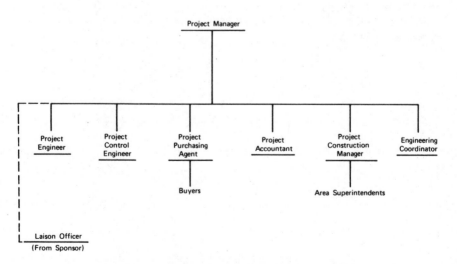

Figure 3.7   Typical construction project taskforce.

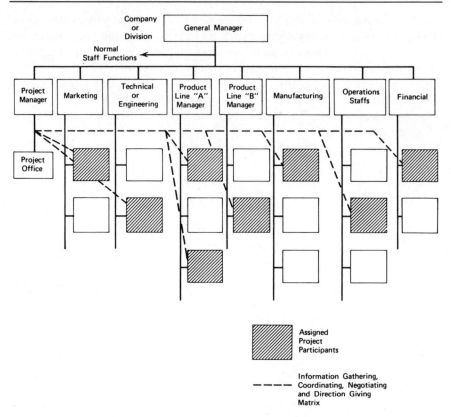

Figure 3.8   Generalized organization chart with project responsibility delegated to full time project manager. The matrix relationships now include direction giving information to carry out decisions of the project manager, in addition to the communicating and coordinating functions.

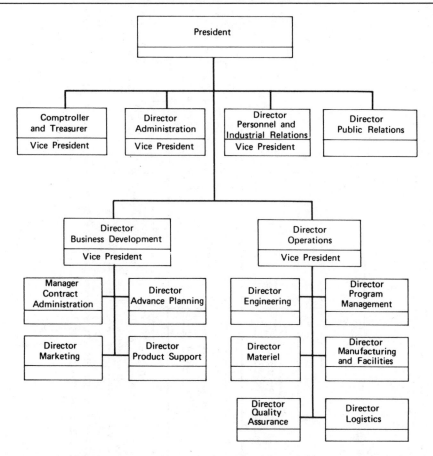

Figure 3.9   Organization of company with a Director, Program Management. Project managers report to this director and have matrix relationships with the other functional departments.

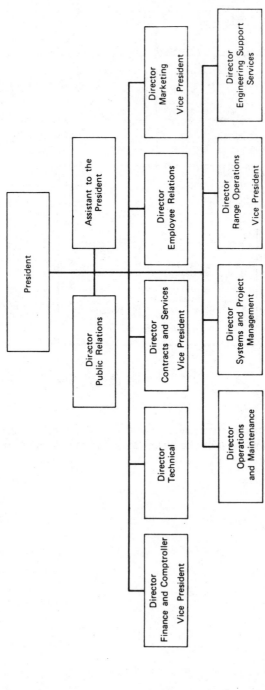

Figure 3.10 Organization of company with a Director, Systems and Project Management. Project managers report to this director and have matrix relationships with the other functional departments.

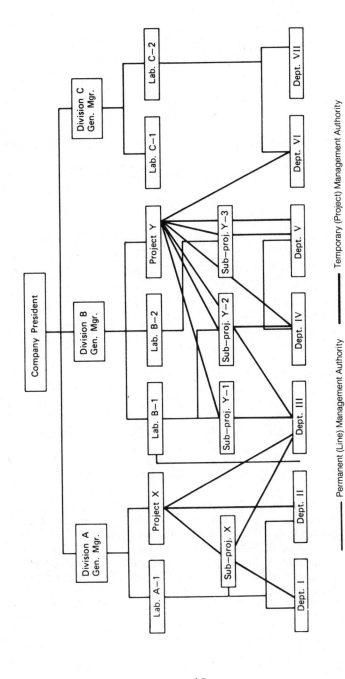

Figure 3.11 Generalized organization for a multidivisional, multiproject situation.

———— Permanent (Line) Management Authority ———— Temporary (Project) Management Authority

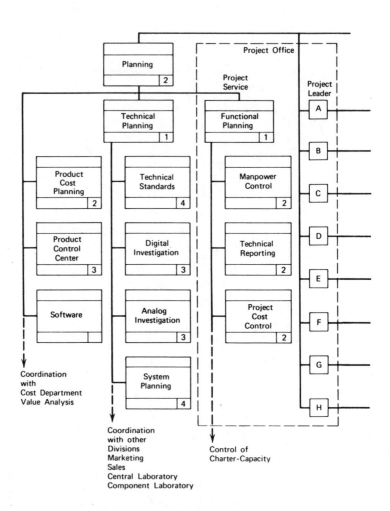

Figure 3.12 Organization of a development department with assigned project leaders. All project leaders report to the Chief Engineer and have matrix relationships with the assigned circuit design, laboratory, and mechanical design engineers. If the projects also require support from marketing, manufacturing, field operations, and product line management, these project leaders would relate to higher level project managers on a matrix basis.

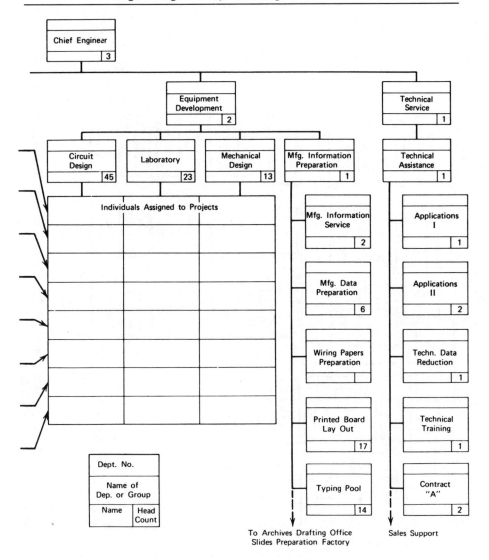

Chief Engineer | 3

Equipment Development | 2

Technical Service | 1

Circuit Design | 45

Laboratory | 23

Mechanical Design | 13

Mfg. Information Preparation | 1

Technical Assistance | 1

Individuals Assigned to Projects

Mfg. Information Service | 2

Applications I | 1

Mfg. Data Preparation | 6

Applications II | 2

Wiring Papers Preparation

Techn. Data Reduction | 1

Printed Board Lay Out | 17

Technical Training | 1

Dept. No.

Name of Dep. or Group

Name | Head Count

Typing Pool | 14

Contract "A" | 2

To Archives Drafting Office
Slides Preparation Factory

Sales Support

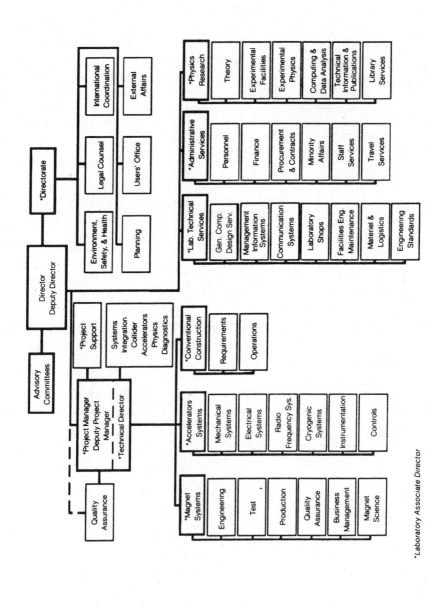

Figure 3.13 Organization Chart for the Superconducting Super Collider (SSC) Laboratory. (Source: Superconducting Super Collider Laboratory, Universities Research Association. Used by permission.)

*Laboratory Associate Director

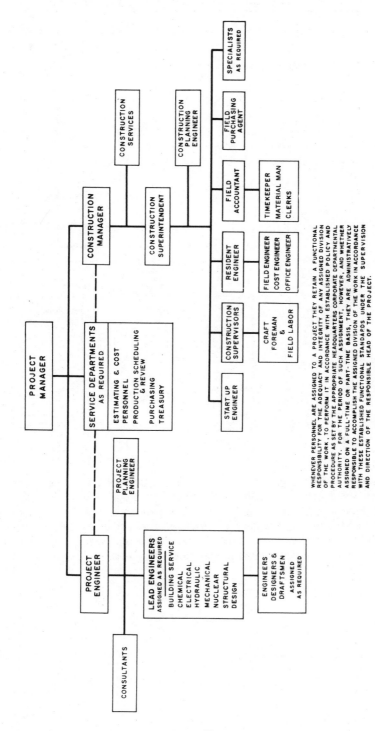

WHENEVER PERSONNEL ARE ASSIGNED TO A PROJECT THEY RETAIN A FUNCTIONAL RESPONSIBILITY FOR THE ADEQUACY AND INTEGRITY OF ANY ASSIGNED DIVISION OF THE WORK, TO PERFORM IT IN ACCORDANCE WITH ESTABLISHED POLICY AND PROCEDURE AS SET BY THE APPROPRIATE HEADQUARTERS CORPORATE DEPARTMENTAL AUTHORITY. FOR THE PERIOD OF SUCH ASSIGNMENT, HOWEVER, AND WHETHER ASSIGNED ON A FULL-TIME OR PART-TIME BASIS, THEY ARE ADMINISTRATIVELY RESPONSIBLE TO ACCOMPLISH THE ASSIGNED DIVISION OF THE WORK IN ACCORDANCE WITH THESE ESTABLISHED FUNCTIONAL STANDARDS UNDER THE SUPERVISION AND DIRECTION OF THE RESPONSIBLE HEAD OF THE PROJECT.

Figure 3.14   Typical organization chart for a large engineering-construction company.

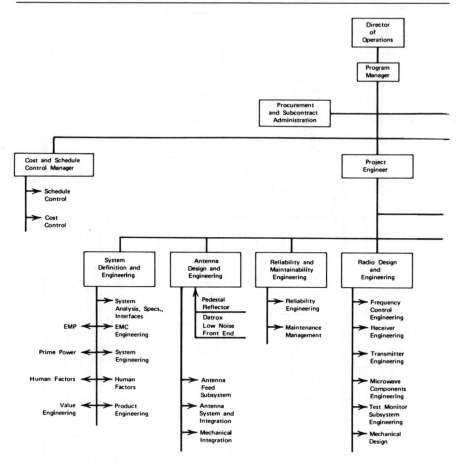

3.15   Program organization chart. This chart is typical of those submitted to a customer to emphasize the strength of the company's program/project management approach. However, in most cases, few of the people represented on the chart actually report to the program or project manager. The lines on such a chart must be clearly understood to represent project-related direction only, with the team members reporting on a "hard-line" basis to their functional managers.

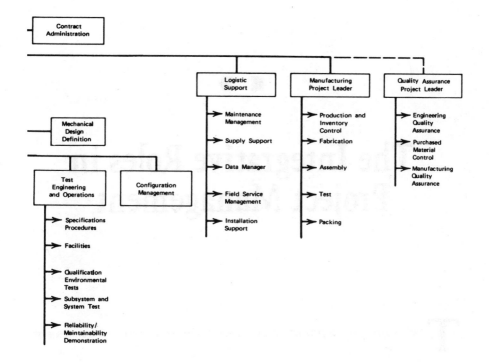

4

《 》

The Integrative Roles in Project Management

The project management triad introduced in Chapter 1 consists of:

- Identified points of integrative project responsibility,
- Integrative and predictive project planning and control systems, and
- The project team: Identifying and managing the project team to integrate the efforts of all contributors to the project.

In this chapter, the first of these project management concepts is discussed.

4.1 THE KEY INTEGRATIVE ROLES

The role of the project manager is a central one, and in fact this role has received considerable attention in the project management literature over the past several decades. However, there are other key integrative roles in project management, and these have frequently been ignored. The key integrative roles are:

- The executive level:
  - The General Manager
  - The Project Sponsor
- The multiproject level:
  - The Manager of Project Management/Multiproject Manager
- The project level:
  - The Project Manager
- The functional department or project contributor level:
  - The Functional Project Leader.

Other important roles may also be identified, including those of the project sponsor (discussed below), the project customer (the person or organization who has authorized the project), the project champion (the person who promotes and keeps the project alive, who may or may not be the general manager), the owner of the results of the project (who may or may not be the project customer), and the user or operator of the project results (who may or may not be the owner). While all of these roles are important, they do not carry the same level of *integrative* responsibility as the key roles.

4.2 GENERAL MANAGER

The role of the general manager (a title used here to denote the person with overall responsibility for a multifunctional division or an entire company) in project management is focused on the organization's overall project management process, and how it is integrated with all other aspects of the organization. Except in rare circumstances, the general manager is not involved directly with the planning and execution of any one project. Rather, this person (1) assures that the right projects are selected or created, (2) provides adequate resources to carry out the approved projects, (3) assures that the project management practices in use within the organization are appropriate, (4) monitors the overall performance on projects, and (5) integrates that performance with the other operations of the organization. In smaller organizations or in organizations with relatively few projects under way, the general manager may act as the project sponsor.

4.3 PROJECT SPONSOR

The project sponsor role is usually held by a senior manager, or a "plural executive" in the form of a steering group or committee, acting for top management of the sponsoring or owner organization. This role may be held by the general manager of the organization responsible for the project, by a higher level executive, or it may be delegated to someone who reports to the general manager. In some cases, the project sponsor role is held by a steering group comprised of key people from various parts of the organization. Only in recent years has the importance of this role been recognized, together with the importance of formally identifying who is assigned to this role for a specific project.

A team of experienced project management professionals concluded that "The absence of a specifically assigned project sponsor with well-defined and understood responsibilities is the cause of many difficulties for projects and project managers. By focusing more attention on this vital role, the effectiveness of project management practices could be improved in most organizations."[1] When these integrative responsibilities have not been assigned to a project sponsor, the project manager often has some difficulty in knowing who to go to for the required decisions. When no project sponsor is identified, some managers (including the project manager at times) may assume that the project manager should act also as the project sponsor, resulting in problems and conflicts caused by reaching beyond assigned authority. Although it is possible to assign one person to both the roles of project sponsor and project manager, in most circumstances these roles should be separated.

4.4 MANAGER OF PROJECT MANAGEMENT

The role of manager of project management is important in organizations in which a significant segment of the business is in the form of projects (see Chapter 1). This position is a recognition of the project management function as an important capability of the organization, along with the more traditional functions of marketing, engineering, procurement, manufacturing, construction/field operations, finance and accounting, legal, and so on. The manager of project management may also be the project sponsor for specific projects. In the absence of a manager of project management, the project sponsor, acting for the general or a higher manager, will give project direction to the project manager.

4.5 PROJECT AND MULTIPROJECT MANAGER

The project manager role is more operational in nature than the more strategic role of the project sponsor. In most situations, the project manager plans and directs the execution of the project to meet the time, cost, and performance objectives as established by the project sponsor. The project manager integrates the efforts of all persons and organizations contributing to the project. The multiproject manager performs the duties of the project manager on several projects at the same time. This role is discussed in more detail in Chapter 8.

A project requires someone to plan, organize, staff, evaluate, direct, control, and lead it from inception to completion. This is defined as the basic *role* of the project manager. This role may be assigned to one person or it may be divided among several people as discussed later.

Primary Responsibilities

- To produce the specific end product or result within the technical, cost, and schedule specifications, and with the organizational resources available.
- To meet the profit objectives of the project, when it is under contract with a customer.
- To integrate the efforts of all project contributors, and provide active leadership to the project team.
- To alert higher management (the project sponsor) at any time that it appears the technical, cost, or schedule objectives will not be met.
- To make or force the required decisions to assure that the project objectives will be met.
- To recommend termination of the project or alternative solution if the project objectives cannot be achieved and contractual obligations permit.
- To serve as the prime point of contact for the project with the customer, top management, and functional managers.
- To negotiate "contracts" (work orders) with the various functional departments for performance of work packages to specification and within time and cost limits.

A more detailed description of the project manager's duties and responsibilities is given in Chapter 9.

4.6 FUNCTIONAL PROJECT LEADERS

On any given project, there will be several functional project leaders whose role is to integrate the project work within their particular functions or subfunctions (marketing, engineering, test operations, manufacturing, production, and so on). The functional project leaders integrate the work being done and the activities of the project team members within their specific function. The project manager integrates the work of all functions at the project level, and the project sponsor integrates and directs the project at the executive level. Functional managers integrate the work on multiprojects within their functions through their day-to-day direction of the functional project leaders. The manager of project management and the general manager integrate the efforts at the multiproject executive level.

4.7   RESPONSIBILITIES AND AUTHORITY

Responsibilities for initiating, planning, executing, controlling, and terminating projects are divided among the persons discussed in the preceding sections. The manner in which these responsibilities are divided and delegated will depend on:

- The size and nature of the parent organization.
- The size and nature of a given project and its current life-cycle phase.
- The number of projects under way.
- The capabilities of the managers involved.
- The maturity of the project management function within the organization.

Responsibilities Retained by the General Manager

Certain project responsibilities are not usually delegated by the general manager (except perhaps to a manager of project management) where major projects are involved. Generally, these include:

- Resolution of project-related conflicts involving senior managers.
- Evaluation of project performance of functional department managers and project managers.
- Periodic evaluation of progress (major milestone accomplishment, forecast of cost and profit at completion, etc.)

Financial and budgetary approvals as specified by pertinent policies and procedures must be adhered to at each level as well as those affecting decisions to bid, and review and approval of contracts.

Project Sponsor Responsibilities

The project sponsor is the focal point for project decisions that are beyond the scope of authority of the project manager. Typical responsibilities of the project sponsor are:[2]

- To hold accountability for the project investment.
- To define and make the business case for the project.
- To approve the project scope and objectives, including schedule and budget.
- To issue project directives as appropriate.
- To appoint the project manager and approve that person's organizational charter and reporting location.
- To monitor the project environment.
- To make and approve project changes and decisions on project requirements.
- To review progress and provide strategic direction to the project manager.
- To set strategic priorities and resolve conflicts escalated by the project manager or other project team members.

Manager of Projects Responsibilities

The manager of project management is typically responsible for (1) providing professional direction and training to the project and multiproject managers; (2) developing and improving the organization's project management process, procedures, and practices; (3) providing project planning, scheduling, estimating, monitoring, and reporting assistance to the project and multiproject managers; and (4) resolving interproject conflicts, consistent with the scope of authority delegated to this position and to the project sponsor by the general manager.

Project Manager Responsibilities

The general manager (or other senior manager who has full responsibility for the project) may delegate very limited or very broad responsibilities to one person on a particular project. If he or she delegates very limited responsibilities, then the general manager retains the actual role of the project manager.

In accordance with generally accepted practice, the project manager in charge of a major project is delegated the basic responsibility of overall direction and coordination of the project through all its phases to achieve the desired results within the established budget and schedule.

In effect, the project manager is the general manager of the project in terms of responsibility, accountability for the final profit or loss on the project, and for meeting the established completion date. The project manager's primary task is the integration of the efforts of all persons contributing to the project. This responsibility does not supplant the responsibility of each functional manager contributing to the project, but rather overlaps the functional responsibilities, with emphasis on the total project.

This basic responsibility may be defined in much greater detail, with specific reference to the areas of planning, scheduling, negotiating, communicating, evaluating, controlling, decision making, and reporting.

Examples of project manager job descriptions are in Chapter 9. In practice, such job specifications must be tailored to the specific project. Differences will occur for each of the basic categories of projects (commercial; product development; research, development, and engineering; capital facilities; systems; and management projects), as well as those differences caused by the nature of the parent organization and the project itself.

Functional Responsibilities on Projects

Three levels of responsibility exist within each functional department contributing to projects.

The functional manager.   This person holds overall responsibility for planning and executing the specific tasks to be performed by his department for each project. The basic specifications of each work package task (result to be achieved, schedule, budget) must be established in a negotiating process between the project manager and the functional manager, or his representative (see below). Within the limits of these specifications, the functional manager has the responsibility of detailed planning, functional policy and procedure direction, functional quality, and providing and developing an adequately skilled staff.

The functional project representative or leader.   For each specific project, this person acts for the functional manager and at the same time represents the project manager within the functional department. He or she is the liaison between the project and the department and is the focal point for all activity on that project within that department.

The work package task manager or leader.   This manager is assigned specific, direct responsibility for one or more tasks related to a project.

Figure 4.1 portrays a generic model of a matrix organization arrangement. This model reflects the integration of the functional managers and the project manager in their *interdependent and complementary* roles. Each of the five key managers are shown with their relationships and basic responsibilities.

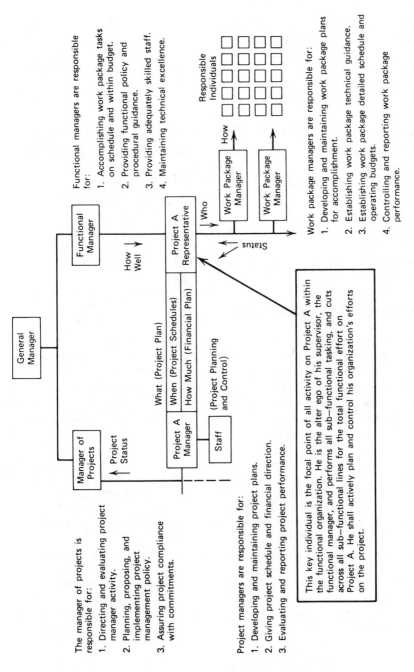

The manager of projects is responsible for:

1. Directing and evaluating project manager activity.
2. Planning, proposing, and implementing project management policy.
3. Assuring project compliance with commitments.

Project managers are responsible for:

1. Developing and maintaining project plans.
2. Giving project schedule and financial direction.
3. Evaluating and reporting project performance.

This key individual is the focal point of all activity on Project A within the functional organization. He is the alter ego of his supervisor, the functional manager, and performs all sub—functional tasking, and cuts across all sub—functional lines for the total functional effort on Project A. He shall actively plan and control his organization's efforts on the project.

Functional managers are responsible for:

1. Accomplishing work package tasks on schedule and within budget.
2. Providing functional policy and procedural guidance.
3. Providing adequately skilled staff.
4. Maintaining technical excellence.

Work package managers are responsible for:

1. Developing and maintaining work package plans for accomplishment.
2. Establishing work package technical guidance.
3. Establishing work package detailed schedule and operating budgets.
4. Controlling and reporting work package performance.

Figure 4.1 Matrix organization of relationship of project management to functional management. (Adapted from Management: A Systems Approach, 3rd Ed., by David I. Cleland and William R. King, Eds., McGraw-Hill, New York, 1983, p. 353.)

Project Authority

Unless the project manager is also the general manager, he will never have the traditional authority commensurate with his responsibility for the total project. To a degree greater than in most other management positions, the project manager's authority stems from his personal abilities to *earn* such authority, in addition to his assigned or *legal* authority. These two sources of authority are summarized:

Legal authority is derived from:	*Earned authority* is derived from:
organizational charter	technical and organizational knowledge
organizational position	management experience
position or job specification	maintenance of rapport
executive rank	negotiation with peers and associates
policy documents	building and maintaining alliances
superior's right to command	project manager's focal position
delegated power	the deliberate conflict
the hierarchical flow	the resolution of conflict
control of funds	being right

Interpersonal Influence Bases

A project manager, or any sort of manager, for that matter, who must elicit performance from others not under his or her direct control must rely on interpersonal influence bases other than formal authority. Wilemon and Gemmill have identified three influence bases that are of major importance to project managers: reward and punishment power; expert power; referent power. The following summary is taken from "Interpersonal Power in Temporary Management Systems."[3]

> *Reward and punishment power* may be either *possessed* or *attributed*. *Possessed* reward and punishment power refers to what a project manager can do to directly or indirectly block or facilitate the attainment of personal goals of people who balk at his requests. It is derived in large part from the charter given to the project manager by his superiors [together with the other factors listed previously as sources of legal authority].
>
> *Attributed* reward and punishment power, on the other hand, refers to the reward and punishment power assigned to the project manager by those with whom he interacts. It is what they believe he can do directly or indirectly to them to block or facilitate their personal goals if they neglect his requests. If the project manager has a direct line to top management, project team members may perceive that he is capable of influencing their careers by comments on their performance, whether or not in fact he has this influence.
>
> While possessed reward and punishment power is usually the basis for attributed reward and punishment power, it is not always present. Differently

put, a project manager has as much reward and punishment power as others perceive him as possessing. If others believe he possesses it when he does not, for all intents and purposes, the effect is the same as if he actually had it. Such bluffing is often effective because the power world of an organization is unofficial and ambiguous and the reward and punishment power of a project manager is more often than not indirect. An interfacing functional manager simply does not ask a project manager how much power he has, or how much influence he has over a third party, or what he can do to him if he fails to do what is requested.

This reward and punishment power may be *direct,* stemming from legal sources, or *indirect,* derived from the earned or personal authority sources. Through negotiation, personality, persuasive ability, competence, reciprocal favors, and the like, a project manager may have a great deal of indirect reward and punishment power, even if he lacks direct reward and punishment power.

Expert power refers to the ability of a project manager to get the contributors to his project to do what he wants them to do because they attribute greater knowledge to him or believe he is more qualified to evaluate the consequences of certain project actions or decisions.

Referent power refers to the responsiveness of project contributors because they are, for some reason or other, personally attracted to the project manager and value both their relationship with him and his opinion of them. Thus, personal friendships and alliances can become an important source of influence for a project manager. If a project manager is personally disliked, he may have negative referent power, which will make the task of influencing the project contributors even more difficult.

Project Manager Selection, Assigned Authority, and Ability to Manage

The general manager must select a project manager with the background and experience that provides a reasonable chance for developing substantial expert and/or referent power with the project contributors. In addition, the general manager must provide the project manager with adequate legal authority to develop the needed amount of reward and punishment power. From there on it is entirely up to the project manager to earn the additional authority from the sources described.

4.8   ALTERNATE WAYS OF FILLING THE PROJECT MANAGER ROLE

The most effective project management results from the assignment of one person to fill the total role of the project manager, and for this person to carry out the responsibilities from the earliest possible moment in the conception of the project to its ultimate completion.

However, for a number of reasons, it is rarely possible to achieve this objective. For example, many more projects pass through the conceptual and definition phases than go on to final completion. Several times as many proposals are usually submitted to potential customers (end of the definition phase) than are accepted and funded by the customers. It is not always possible to assign a project manager to every proposal and know that this person will be available to carry through the project if the award is made. Also, when projects have a relatively long duration (more than 18 months), the needs of the organization (and of the person) often demand that the project manager be promoted to larger or different responsibilities before the project is completed. And sometimes because of the lack of a clear understanding or acceptance of the requirements of the project manager, this role is often arbitrarily or unknowingly divided among various individuals who may be scattered throughout the organization. The most commonly encountered difficulties in this regard, and the most effective solutions, are discussed in the following paragraphs.

Continuity of the Project Manager Assignment

The effectiveness of a project manager is directly related to the continuity of responsibility through the life-cycle of the project. The difficulties have previously been discussed. Possible steps to minimize these difficulties include the following:

- As early as possible, alert the potential project manager to an embryo project and have that person monitor the activity as time permits until a firm assignment is feasible.
- Where a project management department exists, assign a project manager to represent the department on all major proposals. This person will at least evaluate the proposals from the project management viewpoint, and will also provide needed background information to the specific project manager assigned to a contract as it is awarded.
- When selecting a project manager, establish a clear understanding with the selectee and upper management levels that the person will be expected to remain in the assignment until the project is complete.
- Avoid the practice of "passing the baton" for project responsibility from department to department as a project passes from phase to phase. If the project manager is actually transferred from department to department, then the continuity of personal responsibility is preserved.

- Establish, through the Project File (see Chapter 10), a well-documented history, including a set of plans and performance against the plans, to become the "turnover file" should it be necessary to change the project manager prior to completion.

- Attempt to overlap project assignments, where possible, by assigning a project manager whose project is nearing completion, and therefore may demand less time, to a new project which can be followed part-time during the early conceptual phase and then on a full-time basis when the first project is complete.

Full-Time versus Part-Time Assignment

Large projects will require one person to be assigned full-time to the role of project manager. Smaller projects do not require and cannot justify such a full-time assignment, but the need remains to fill the role, nevertheless, in many cases.

Several commonly encountered causes of ineffective project management relate to the part-time assignment of project managers. Frequently the required number of project managers are not readily available, and various compromises result.

Projects Assigned to Functional Managers

A manager who has a full-time functional responsibility may be given the added burden of acting as the project manager for one or more projects. This is generally ineffective, unless 80 to 90 percent of the work to be done on the project will be performed by people reporting to this manager through the organization structure.

Either the functional or the project responsibility will suffer simply due to lack of time. Or, the manager's primary allegiance is to the department and not the project. When conflicts occur, the project will probably suffer.

Also, when counterpart functional managers are enlisted to aid on a project, they are caught between the realization that if they help to make the project very successful, they are bolstering the chances of a rival being promoted instead of themselves; and that if, on the other hand, the project can be caused to fail at least partially, then the competitive edge of the manager running the project has been reduced. When this part-time assignment is combined with the "pass-the-baton" approach previously discussed, rivalry is intensified by the (sometimes very justified) feeling that each manager is passing along many unsolved problems. The last person to receive the project assignment may have inherited an impossible task, and it is usually very difficult to pinpoint the person responsible for failure of the project.

Finally, because the characteristics of projects are different from those of a functional organization, as discussed in Chapter 2, most managers find it difficult to switch alternatively every day from managing a project to managing a function. This can create unusually high stress in the manager and the project team members.

Several Projects Assigned to One Full-Time Project Manager

A more effective solution in situations where not enough project managers are available for full-time assignment one to a project, or where a number of projects exist, none requiring a full-time manager, is to assign several projects to one full-time project manager. This approach has the advantage that the person is continually acting in the same role, that of a project manager, and is not distracted or encumbered by functional responsibilities. His primary allegiance is to his projects, and his skills as a project manager can be more rapidly developed.

In the process of planning, controlling, evaluating, and directing one project, the project manager can frequently cover other projects when in contact with functional managers and team leaders, thus minimizing the time expended by the functional managers for this purpose. Interproject priorities and conflicts can often be resolved directly by the project manager for his assigned projects. Rivalry between functional managers is reduced.

Division of the Project Manager's Responsibilities

The responsibilities of the project manager are frequently divided among several persons. It is sometimes difficult to answer the question, "Who is the real project manager?" Usually the real project manager is found to be the person to whom all those carrying out portions of the project manager role report. It is often a surprise to this manager to discover that he is, in fact, the project manager, as he may be several levels up the reporting structure.

The most frequently found method of dividing project responsibilities is to assign one person the technical (product) responsibilities, another the schedule, and a third the cost. Many times a fourth person holds the contract administration responsibilities, perhaps another serves as the prime customer contact point, and still another is concerned with the manufacturing aspects.

Such division of the project manager's responsibilities is probably the most common cause of projects not achieving their objectives. Unless one person integrates the efforts of the project engineer, the project planner and controller, the project cost engineer, the project contract administrator, and so on, it is not possible to evaluate the project effectively to identify current or future problems and initiate corrective action in time to assure that the objectives will be met.

The project manager cannot actually perform all of the planning, controlling, and evaluation activities needed, any more than he can perform all of the technical specialty activities required. Project management support services must be provided, and the project manager must direct and control these support activities. The hazard is that the support activities may exist, but in the absence of an assigned project manager, they are not properly used.

Retention of the Project Manager Role by the General Manager

In certain situations, such as with one major project of extreme importance to the Company or Division, the general manager may properly elect to retain the role of project manager. This may also be appropriate with multiple small projects, as discussed in Chapter 8. In such cases it is usually desirable to appoint a project coordinator to perform much of the planning and communicating related to the project for the general manager. The project coordinator should report directly to the general manager in this role and should integrate the needed project management support functions.

Filling the Project Manager Role in Multiple Project Situations

A number of organizations are responsible for few major projects but many smaller ones. In such situations the project manager role may remain with the appropriate level of line management, but responsibility must be clearly established for integrated planning and predictive control of all projects to support and coordinate the efforts of all functional managers. This responsibility may be assigned to a position with various titles such as planning manager, operations control manager, projects coordinator, projects controller, and so on.

As discussed in Chapter 8 and elsewhere in this book, such integrative planning and control elements of project management are very important.

4.9 CHARACTERISTICS, SOURCES, AND SELECTION OF PROJECT MANAGERS

The effectiveness of a project manager depends heavily on his or her personal experience and characteristics, more so than in many other management positions below the general manager level. The ability to function effectively in a relatively unstructured relationship to other managers, to earn authority, to integrate the efforts of many people, and to resolve conflicts appropriately are very important to the project manager's success.

Key Personal Characteristics

Some of the key personal characteristics of effective project managers are:

- Flexibility and adaptability.
- Preference for significant initiative and leadership.
- Aggressiveness, confidence, persuasiveness, verbal fluency.
- Ambition, activity, forcefulness.
- Effectiveness as a communicator and integrator.
- Broad scope of personal interests.
- Poise, enthusiasm, imagination, spontaneity.
- Able to balance technical solutions with time, cost, and human factors.
- Well organized and disciplined.
- A generalist rather than a specialist.
- Able and willing to devote most of their time to planning and controlling.
- Able to identify problems.
- Willing to make decisions.
- Able to maintain proper balance in the use of their time.

The Project Manager's Skills

A team of experienced project management professionals developed a profile of the project manager's skills and knowledge and concluded that *the project manager's integrative skills are the most important for project success.*[4] Figure 4.2 shows a summary profile of the project manager and the central importance of the integrative skills. The integrative skills of most importance and use to the project manager include:

- Holistic thinking.
- Using a systems approach.
- Being flexible, adaptable, open-minded.
- Ability to set and balance priorities.
- Cross-cultural abilities (macro and micro).

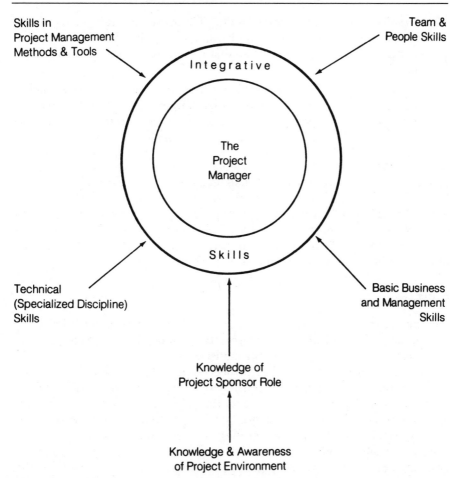

Figure 4.2   Project manager skills profile. (Source: Russell D. Archibald, and Alan Harpham, "Project Managers Profiles and Certification Workshop Report," *Proceedings of the 14th International Expert Seminar,* March 15–17, 1990, of the INTERNET International Project Management Association, Zurich, p. 10. Used by permission.)

The remaining four skill areas shown in Figure 4.2 are:

- *Skills in project management methods and tools:* These represent the classic practices in project management that are described throughout the book, and elsewhere in the project management literature. These relate to planning, organizing, monitoring, and controlling the project.
- *Team and people skills:* These include the interpersonal skills that are required to lead, communicate, coordinate, facilitate, motivate, and build the project team.

- *Technical (specialized discipline) skills:* These are the engineering, scientific, economic, mathematical, or other skills related to the particular discipline that forms the predominant experience bank of the project manager.
- *Basic business and management skills:* This area covers the basic understanding of how the business or industry operates, how companies and other organizations are managed, and fundamental methods of planning, budgeting, financing and operating organizations.

A certain minimum level of knowledge and skill in the last two areas, technical and basic business and management, are considered to be prerequisites for any person to be considered for the job of project manager. The other three skill areas—team and people, project management methods and tools, and integration—are often learned on the job. Identification of these five skill areas can be useful in structuring the subject matter to be used in training project managers and their teams.

Sources

The prime sources of project managers (either within or outside of the parent organization) are: other projects, product line managers, functional managers, and functional specialists.

The best source is from other projects—provided the projects and the project managers have been successful. However, there is an ever-increasing demand for project managers, and it is very difficult to find a project manager who has just completed a project at the right time for each new project.

Functional managers are sometimes given a full-time project manager assignment. If someone has been a first level or middle manager, the project environment may be extremely strange to him, and unless he or she has the proper personality traits and is provided with proper training and guidance, they may have a difficult time carrying out their responsibilities.

Functional specialists are often given a project manager assignment and may become quite effective in such an assignment. The greatest difficulty usually relates to the changeover from *doing* a specific type of work to *managing* the efforts of others. It is necessary, to be most effective, for the specialist to become a generalist and not try to continue his specialty on a major project for which he is also the project manager. On small projects this is often unavoidable, and requires good understanding of the differences between the two roles.

Selection

Selection of a person to be a project manager should not be a unilateral procedure performed by the general manager or other senior manager.

The position and the project should be described frankly to the candidate, who should be given the freedom to accept or reject it without prejudice. Unless he is personally motivated by the opportunity and the nature of the challenge, he will probably not become an effective project manager. Every effort should be made to select a person who has a reasonable chance to stay with the project to its completion.

4.10 CAREER DEVELOPMENT IN PROJECT MANAGEMENT

As recognition grows of the nature and importance of project manager assignments, the need becomes more apparent for more formalized career development in project management. Because projects begin and end, project management assignments are less secure than functional assignments. Frequently, the best qualified people cannot be attracted to projects, because they must leave the known security of a functional department for an unknown future when the project ends.

In an increasing number of organizations, the project management function has been established as a part of the fairly permanent organizational structure. This provides a base on which to build continuity of project assignments, long-term security for project-oriented employees, and more effective career development programs. Examples of such organizational structures are shown in Chapter 3.

Development

The integrative skills required by project managers can best be developed through actual experience on projects. However, this development can be accelerated, and the effectiveness of the manager increased, through appropriate development and training seminars and workshops, and through direct exposure to the methods employed by other successful project managers. Two seminar outline examples are given in the Appendices.

Since projects exist in all sizes, it is possible to assign a less experienced person to a smaller or less complex project. If successful with that, the person can then be assigned to larger projects.

Performance Evaluation and Career Planning

Evaluation of a project manager's performance is often more difficult than evaluating other managers. This is because so many factors affect the project over which the project manager has little direct (legal) authority and control.

A highly successful project may or may not be the direct result of the project manager's efforts, just as an unsuccessful project may have had the best project manager in the organization in charge of it. At this stage in the evolution of the project management approach, no formalized or systematic method of evaluating project manager performance has been developed, but obviously the overall project results do count heavily in this evaluation.

Another area of potential difficulty is the follow-on assignments of project managers, and the affect on their careers of a two or three year project manager assignment. Highly successful project managers may be in an even more difficult position, since top level positions commensurate with their abilities are less frequently in need of filling. Top management attention must be given to this problem area if the best qualified persons are to be attracted to the project manager assignment.

Two factors provide possible solutions to this problem. First, the number of projects and their recognition are growing, hence the demand for project managers is increasing. Flexibility in assignment of project managers between divisions and areas, assuming similar kinds of projects and no substantial language problems, can also be of assistance in this regard, since one division having a capable, unassigned project manager can allow transfer to another in need of such a manager.

Second, the consolidation of project management functions within the organization, as appropriate to the environmental conditions, can provide assistance in the overlapping of project assignments, the development of project manager skills, and the continuity of future assignments.

Recognition of project management as a possible career assignment and provision for planned professional growth in this area will further assist in attracting the needed capabilities. The similarities between the project manager job and higher level positions in general management have led to the realization that project experience is valuable in the development of general management skills.

5

‹ ›

The Project Team and
Key Human Aspects of
Project Management

Recognition that projects are planned and executed through the integrated efforts of a diverse group of people—namely, the project team—and then getting those people to actually function as a team is a fundamental concept underlying effective project management. Together with the identification of the integrative roles discussed in Chapter 4, and the integrative and predictive planning and control systems discussed in later chapters, the principles of the project team and teamworking form the triad of concepts that differentiates project management from other types or forms of management. While teamwork is not unique to project management, it is a fundamental requirement for effective management of projects.

In this chapter, the conditions for effective project teamwork are discussed, and a few of the key human aspects of special importance in project management are presented. There are many important topics in the area of interpersonal relations and the management of human resources (communications, negotiations, personal time management, decision making and problem solving, motivation, supervisory skills, leadership, and so on) that are not included here. A number of excellent books are available on these subjects.

5.1 THE PROJECT TEAM CONCEPT

A project is comprised of a number of diverse tasks. Therefore different people—each having the required expertise and experience—are needed to perform each task. In the broadest sense, all persons contributing to a project are members of the project team. On larger projects where several hundred or even several thousand people are working, we must identify the *key* project team members. These key team members will include the project manager (the team leader), the functional project leaders, and the lead project support people. More specific identification of the key project team members is given in Chapter 9.

The term "functional project leader" is used here generically, and includes people within the project's parent organization as well as people in outside organizations, such as consultants, contractors, vendors, and suppliers. If a project is large enough to be broken into subprojects, then the team concept applies equally to those subprojects. The overall project team would include the subproject manager (or functional project leader) as a team member. That person would also be the leader of the subproject team. In many projects, the client or customer is an active contributor and therefore is included as a member of the team. When possible, inclusion on the project team of representatives of other outside organizations that contribute in some way to the project can be very beneficial. Such organizations include financial institutions, regulatory or oversight agencies, and labor unions.

To have an effective project team, as distinct from simply a group of people working on loosely related tasks, several conditions are necessary:

- Identification of the project team members and definition of the role and responsibilities of each.
- Clearly stated and understood project objectives.
- An achievable project plan and schedule.
- Reasonable rules (procedures regarding information flow, communication, team meetings, and the like).
- Leadership by the project manager.

If any of these conditions is not present it will be difficult to achieve effective teamwork.

5.2 EFFECTIVE TEAMWORKING

To have an effective team, the team players must be identified. Project managers often fail to do this, or they may only identify their team members on an "as needed" basis when a new task comes up that cannot be performed by someone already on the team. In some cases, the project manager may know the team members, but will fail to inform the other members, so only the project manager knows who is on the team.

A useful practice in identifying who is on the project team is to start by identifying the *project stakeholders,* or all those persons who have a stake (a vested interest, responsibility, decision power) in the project and its results. Briner et al. have suggested mapping the principal stakeholders and the project team, as shown in Figures 5.1 and 5.2.[1]

After completing the mapping exercise, a listing of all project team members is compiled and distributed to the entire team. This list should include each team member's full name, address (regular and electronic mail), voice and facsimile telephone numbers, and any other pertinent communication information. Frequently, this list will include home telephone numbers. For those project teams that have established escalation procedures (for resolving issues, conflicts, or other problems), the team member's immediate supervisor with office and home telephone numbers are also listed.

The general duties and responsibilities of each team member will normally be documented by the organization's human resource practices, following the pattern shown in the examples in Chapter 9. However, for effective project teamwork, it is imperative to define the responsibilities of each team member for each task to be carried out on their specific project. The best tool available for this purpose is the task/responsibility matrix illustrated in Chapter 10. This must be based on the project/work breakdown structure, also described in Chapter 10.

Clearly Stated and Understood Project Objectives

The basic project objectives will usually be known prior to identifying the project team members. However, a team effort is required to clarify, expand on, and quantify these initial project objectives to produce statements of objectives that all members of the team understand, accept and are committed to. Hastings et al.[2] point out that teams must be aware that there are multiple and often conflicting sets of expectations about their performance on the project, including expectations from outside the project, the team, and each individual team member. These authors suggest thinking about good performance and successful achievement along two dimensions—the hard/soft dimension and the acceptable/excellent dimension. The hard/soft dimension refers to two different kinds of *criteria* of performance, and the acceptable/excellent dimension refers to two different *standards* of performance.

The Hard / Soft Dimension

The hard/soft dimension concerns the tangible and intangible aspects of performance. Hard criteria tend to be measurable, the most frequent being

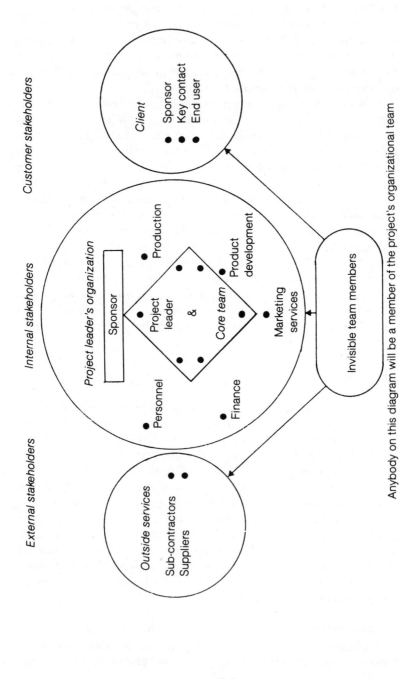

Figure 5.1 Stakeholders in a project. (Source: Wendy Briner, Michael Geddes and Polin Hastings, "Project Leadership," Van Nostrand Reinhold, New York, 1990, p. 41. Used by permission).

Anybody on this diagram will be a member of the project's organizational team

94

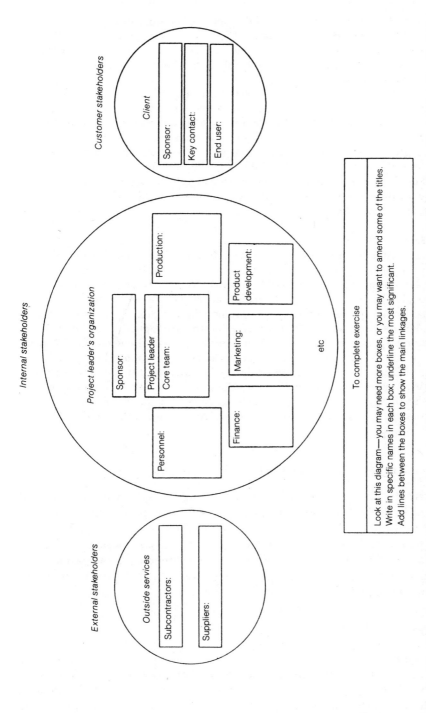

Figure 5.2 Exercise: Mapping your team. (Source: Wendy Briner, Michael Geddes, and Polin Hastings, "Project Leadership," Van Nostrand, Reinhold, New York, 1990, p. 45. Used by permission).

95

to do with time, cost, resources and technical standards. Soft criteria on the other hand are more subjective and difficult to measure. Yet they are clearly used frequently in evaluating performance. They are more about "how" the task was accomplished, the attitudes, skills and behavior demonstrated by the team and its members. . . .

In setting success criteria ordinary teams tend to concentrate on *hard* criteria only and ask questions such as, "How many, how much and when?" Superteams do all this too (and mostly more punctiliously) but add another dimension. They also draw out clients' and sponsors' more subtle expectations, those to do with ways of working and the relationships with the client, to attitudes adopted on such things as quality, reliability and attention to detail. These are all factors that are crucial to a client's ultimate satisfaction. Equally these soft criteria are explored, clarified and agreed with the sponsor, and service departments.

The Acceptable/Excellent Dimension

The acceptable/excellent dimension on the other hand concerns standards of performance. And it is around this dimension that the whole Superteam idea was originally crystallized. In a world where the best is no longer good enough, the frontiers of performance are always being stretched. "The best can always be bettered" could almost be the Superteam motto. We find many teams who think that their performance is good, but who in fact are underperforming. They may be averagely good when compared with those other teams they see. Their performance is acceptable but in no way outstanding. . . . Superteams strive to be different, and achieve just a little bit more than the competition. They are constantly looking for ways to do things better, constantly testing their assumptions about what is achievable and searching for ways to overcome any problems that lie in the path.[3]

An Achievable Project Plan and Schedule

Effective teamwork depends heavily on having a project plan and schedule that reflects the way the team members will actually do the work. The team must understand and be committed to the plan and schedule, which must be reasonably achievable. Chapter 10 presents the basic project planning and scheduling tools that have been developed over the past several decades, and Chapter 11 describes more recently developed methods for setting the stage for effective project teamworking.

Reasonable Rules

Trying to achieve good teamwork on a complex project without having established reasonable rules, procedures and practices for how the project will be planned, the work authorized, progress reported and evaluated, conflicts resolved, and so on, is like collecting the best athletes from six different sports and turning them loose on an open, unmarked field with instructions to "play the game as hard as you can."

Each organization must develop its own set of project procedures covering the topics of importance within its environment. On large projects, such procedures are usually tailored to the specific needs of that project and issued to all team members in the form of a project procedures handbook, manual, or some similar document. The project procedures usually rely on established corporate practices and procedures wherever possible, and avoid duplication or conflict with such practices. Many of the tools, methods, and practices described in this book will be tailored and incorporated into the project procedures.

Leadership by the Project Manager

Extensive literature exists on the subject of leadership, and it is not the intent here to treat this complex and important subject in great detail. The key point to be made is that the project manager is expected to be the *leader* of the project. Successful project managers have used many different styles and methods of leadership, depending on their own personalities, experience, interpersonal skills, and technical competence on the one hand, and the characteristics of the project and its environment on the other. Owens concluded the following regarding project leadership and related behavioral topics:

Leadership behavior. Project managers cannot rely on one particular leadership style to influence other people's behavior. Different situations call for different approaches, and leaders must be sensitive to the unique features of circumstances and personalities.

Motivational techniques. An awareness of unfulfilled needs residing in the team is required to successfully appraise motivational requirements and adjust a job's design to meet those needs.

Interpersonal and organizational communications. Conflict situations occur regularly. A problem-solving or confrontation approach, using informal group sessions, can be a useful resolution strategy.

Decision-making and team-building skills. Participative decision making meets the needs of individual team members and contributes toward effective decisions and team unity.[4]

In their discussion of the project leader as integrator, Briner et al. identify 14 integrative processes that are important to the leader of a project. These are illustrated in Figure 5.3 and more fully explained in their book, *Project Leadership.*[5]

The following sections of this chapter and Chapter 6 describe useful ideas for project managers to improve their leadership skills.

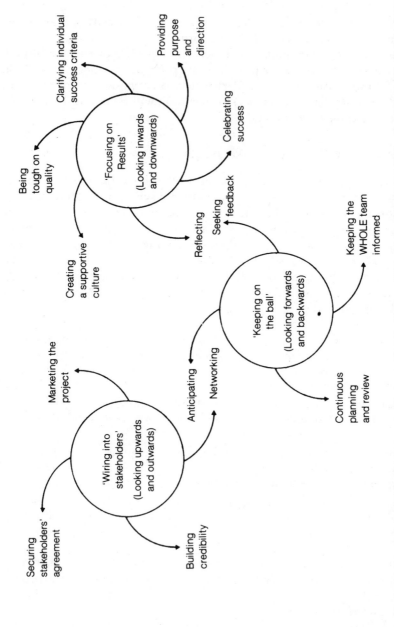

Figure 5.3 Fourteen integrative processes. (Source: Wendy Briner, Michael Geddes, and Colin Hastings, "Project Leadership," Van Nostrand, Reinhold, New York, 1990, p. 19. Used by permission.)

5.3   CONFLICTS AND THEIR RESOLUTION

The results of a research project by Thamhain and Wilemon[6] involving 100 project managers identified seven sources of potential conflict.

1. *Conflict over Project Priorities.* The views of project participants often differ over the sequence of activities and tasks that should be undertaken to achieve successful project completion. Conflict over priorities may occur not only between the project team and other support groups but also within the project team.

2. *Conflict over Administrative Procedures.* A number of managerial and administrative-orientated conflicts may develop over how the project will be managed; that is, the definition of the project manager's reporting relationships, definition of responsibilities, interface relationships, project scope, operational requirements, plan of execution, negotiated work agreements with other groups, and procedures for administrative support.

3. *Conflict over Technical Opinions and Performance Trade-Offs.* In technology-oriented projects, disagreements may arise over technical issues, performance specifications, technical trade-offs, and the means to achieve technical performance.

4. *Conflict over People Resources.* Conflicts may arise around the staffing of the project team with personnel from other functional and staff support areas or may arise from the desire to use another department's personnel for project support even though the personnel remain under the authority of their functional or staff superiors.

5. *Conflict over Cost.* Frequently conflict may develop over cost estimates from support areas regarding various project work breakdown packages. For example, the funds allocated by a project manager to a functional support group might be perceived as insufficient for the support requested.

6. *Conflict over Schedules.* Disagreements may tend to center around the timing, sequencing, and scheduling of project-related tasks.

7. *Personality Conflict.* Disagreements may tend to center on interpersonal differences rather than on "technical" issues. Conflicts often are "ego-centered."

Intensity and Source of Conflict over the Life Cycle of Projects

Figure 5.4 graphically illustrates the relative intensity of conflict from each of the several specific conflict sources for each of four phases in the

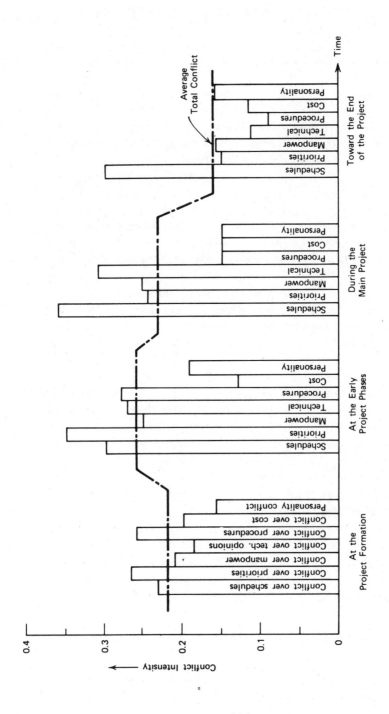

Figure 5.4 Relative intensity of conflict over the life cycles of projects. (Source: Hans J. Thamhain and David L. Wilomen, "Conflict Management in Project Life Cycles," Sloan Management Review, Summer, 1975. Used by permission.)

project life cycle. This indicates that conflict over schedules is strong throughout the project life, while the other conflict sources change perceptibly as the project matures.

As indicated, disagreements over schedules result in the most intense conflict over the total project, relative to other situations. Scheduling conflicts often occur with other support departments over whom the project manager may have limited authority and control. Scheduling problems and conflicts also often involve disagreements and differing perceptions of organizational departmental priorities. For example, an issue urgent to the project manager may receive a low priority treatment from support groups and/or staff personnel because of a different priority structure in the support organization. Conflicts over schedules frequently result from the cumulative effects of other areas involved in project performance, namely, technical problems and manpower resources.

Conflict over project priorities ranked second highest over the project life cycle. In our discussions with project managers, many indicated that this type of conflict frequently develops because the organization did not have prior experience with a current project undertaking. Consequently, the pattern of project priorities may change from the original forecast, necessitating the reallocation of crucial resources and schedules, a process which is often susceptible to intense disagreements and conflicts. Also, in a similar vein to the conflicts over schedules, priority issues often develop into conflict because of changing requirements made to other support departments, hence disturbing their established schedules and work patterns.

Conflict over manpower resources was the third most important source of conflict. Project managers frequently lament that there is little "organizational slack" in terms of manpower resources, a situation in which they often experience intense conflicts. Project managers note that most of the conflicts over personnel resources occur with those departments who either assign personnel to the project or support the project internally.

The fourth strongest source of conflict involved disagreements over technical opinions and trade-offs. Often the groups who support the project are primarily responsible for technical inputs and performance standards. The project manager, on the other hand, is accountable for cost, schedule, and performance objectives. Since support areas are usually responsible for only parts of the project they may not have the broad management overview of the total project. The project manager may, for example, be presented with a technical issue by a support group which involves alternative ways of solving a technical problem. Often he must reject the technical alternative due to cost or schedule restraints. In other cases, he may find that he disagrees with the opinions of others on strictly technical grounds.

Conflict over administrative procedures ranked fifth in the profile of seven conflict sources. It is interesting to note that most of the conflict over administrative procedures occurs almost uniformly distributed with functional departments, project personnel, and the project manager's superior. Examples of conflict originating over administrative issues may involve disagreements over the project manager's authority and responsibilities, reporting relationships, administrative support, status reviews, interorganizational interfacing, etc. For the most part, disagreements over administrative procedures involve issues of how the project manager will function and how he relates to the organization's top management.

Personality conflict ranked low in intensity by the project managers. Our discussions with project managers indicated that while the frequency of personality conflicts may not be as high as some of the other sources of conflict they are some of the most difficult to deal with effectively. Personality issues also may be obscured by communication problems and technical issues. A support person may, for example, stress the technical aspect of a disagreement with the project manager while, in fact, the real issue is a personality conflict.

Cost, like schedules, is often a basic performance measure in project management. As a conflict source, cost ranked lowest. Disagreements over cost frequently develop when project managers negotiate with other departments who will perform subtasks on the project. Project managers with tight budget constraints often want to minimize costs while support groups may want to maximize their part of the project budget. In addition, conflicts may occur as a result of technical problems or schedule slippages which may increase costs.[7]

Conflict Handling Modes

The most and least important modes of conflict resolution, as reported by Thamhaim and Wilemon[8] are summarized in Figure 5.5. The five commonly used modes for handling conflict, as reported by Thamhaim and Wilemon, are:

1. *Confrontation problem solving.* Facing the conflict directly which involves a problem-solving approach whereby affected parties work through their disagreements.

2. *Compromising.* Bargaining and searching for solutions that bring some degree of satisfaction to the parties in a dispute. Characterized by a "give-and-take" attitude.

3. *Smoothing.* De-emphasizing or avoiding areas of difference and emphasizing areas of agreement.

4. *Forcing.* Exerting one's viewpoint at the potential expense of another. Often characterized by competitiveness and a win/lose situation.

5. *Withdrawal.* Retreating or withdrawing from an actual or potential disagreement.

As indicated [in Figure 5.5], confrontation was most frequently utilized as a problem-solving mode. This mode was favored by approximately 70% of the project managers. The compromise approach which is characterized by trade-offs and a give-and-take attitude ranked second, followed by smoothing. Forcing and withdrawal ranked as fourth and fifth most favored resolution modes, respectively.

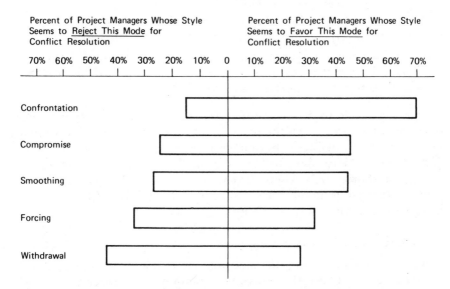

Figure 5.5   Conflict resolution profile. The various modes of conflict resolution actually used to manage conflict in project-oriented work-environments. (Source: Hans J. Thamhain and David L. Wilomen, "Conflict Management in Project Life Cycles," *Sloan Management Review,* Summer, 1975. Used by permission.)

In terms of the most and least favored conflict resolution modes, the project managers had similar rankings for the conflict handling method used between him and his personnel, his superior, and functional support departments except in the cases of confrontation and compromise. While confrontation was the most favored mode for dealing with superiors, compromise was more favored in handling disagreements with functional support departments.[9]

Responsibility Conflicts

Conflict over administrative procedures often results from overlapping responsibilities between project and functional managers. The functional project leader receives direction from two persons, the project manager and the functional superior. This dual responsibility can lead to ambiguities and to increased conflict if the charter of the project manager is not properly defined and understood.

Conversely, responsibility conflicts are minimized if all parties have a clear understanding of the role and responsibilities of the project manager. In an oversimplified manner, the division of responsibilities may be described as:

- *The project manager* provides direction regarding *what* the project tasks are, *when* they should start and finish to meet the overall project goals, and *how much money* is *available* to perform the work.
- *The functional manager* provides direction regarding *who* will perform the tasks, *how* the technical work will be accomplished, and *how much money* is *required* to perform the work.

It is impossible to differentiate these responsibilities sharply. In each case, negotiation between the project and functional managers is usually necessary. There will still be numerous occasions where honest differences can only be resolved by higher authority, but the objective is to call on such authority only when absolutely necessary. The concept of the project manager as an interface manager as described in Chapter 13 has proven to be a useful way to minimize such conflicts.

Implications of Conflict Management

The three areas most likely to cause problems for the project manager over the entire project life cycle are disagreements over schedules, project priorities, and manpower resources. One reason these areas are apt to produce more intense disagreements is that the project manager may have limited control over other areas that have an important impact on these areas, particularly the functional support departments. These three areas (schedules, project priorities, and manpower resources) require careful surveillance throughout the life cycle of a project. To minimize the detrimental conflict, intensive planning prior to actually launching the project is recommended. Planning can help the project manager anticipate many potential sources of conflict before they occur. Scheduling, priority setting, and resource allocation require effective planning to avoid problems later in the project. In our discussions with project managers who have experienced problems in these areas, almost all maintain that these problems frequently originate from lack of effective preproject planning.

Managing projects involves managing change. It is not our intention to suggest that all such problems can be minimized. There always will be random, unpredictable situations which defy forecasting in project environments.

Some specific suggestions are summarized in Figure 5.6. The table provides an aid to project managers in recognizing some of the most important sources of conflict which are most likely to occur in various phases of projects. The table also suggests strategies for minimizing their detrimental consequences.

As one views the seven potential sources of disagreements over the life of a project, the dynamic nature of each conflict source is revealed. Frequently, areas which are most likely to foster disagreements early in a project, become less likely to induce severe conflicts in the maturation of a project.

Project Life Cycle Phase	Major Conflict Source and Recommendations for Minimizing Dysfunctional Consequences	
	CONFLICT SOURCE	RECOMMENDATIONS
Project formation	Priorities	Clearly defined plans. Joint decision-making and/or consultation with affected parties.
		Stress importance of project to organization goals.
	Procedures	Develop detailed administrative operating procedures to be followed in conduct of project. Secure approval from key administrators. Develop statement of understanding or charter.
	Schedules	Develop schedule commitments in advance of actual project commencement.
		Forecast other departmental priorities and possible impact on project.
Buildup phase	Priorities	Provide effective feedback to support areas on forecasted project plans and needs via status review sessions.
	Schedules	Schedule work breakdown packages (project subunits) in cooperation with functional groups.
	Procedures	Contingency planning on key administrative issues.
Main program	Schedules	Continually monitor work in progress. Communicate results to affected parties.
		Forecast problems and consider alternatives.
		Identify potential "trouble spots" needing closer surveillance.
	Technical	Early resolution of technical problems.
		Communication of schedule and budget restraints to technical personnel.
		Emphasize adequate, early technical testing.
		Facilitate early agreement on final designs.
	Manpower	Forecast and communicate manpower requirements early.
		Establish manpower requirements and priorities with functional and staff groups.
Phaseout	Schedules	Close schedule monitoring in project life cycle.
		Consider reallocation of available manpower to critical project areas prone to schedule slippages.
		Attain prompt resolution of technical issues which may impact schedules.
	Personality and manpower	Develop plans for reallocation of manpower upon project completion.
		Maintain harmonious working relationships with project team and support groups. Try to loosen up "high-stress" environment.

Figure 5.6 Major conflict sources by project life cycle stages. (Source: Hans J. Thamhain and David L. Wilomen, "Conflict Management in Project Life Cycles," *Sloan Management Review,* Summer, 1975. Used by permission.)

Administrative procedures, for example, continually lose importance as an intense source of conflict during project maturation. By contrast, personality conflict, which ranks lowest in the project formation stage, is the second most important source of conflict in project phase out. In summary, it is posited that if project managers are aware of the importance of each potential conflict source by project life cycle, then more effective conflict minimization strategies can be developed.

In terms of the means by which project managers handle conflicts and disagreements, research data reveal that the confrontation or problem-solving mode was the most frequent method utilized. While these studies do not attempt to explore the effectiveness of each mode separately, earlier research by Burke[10] suggests that the confrontation approach is the most effective conflict-handling mode.[11]

In some contrast to studies of general management, research findings in project-oriented environments suggest that it is less important to search for a best mode of effective conflict management under all conditions. It appears to be more significant that project managers, in their capacity of integrators of diverse organizational resources, employ the full range of conflict-resolution modes. While confrontation was found as the ideal approach under most circumstances, other approaches may be equally effective depending upon the situational content of the disagreement. Withdrawal, for example, may be used effectively as a temporary measure until new information can be sought or to "cool-off" a hostile reaction from a colleague. As a basic long-term strategy, however, withdrawal may actually escalate a disagreement if no resolution is eventually sought.

In other cases, compromise and smoothing might be considered an effective strategy by the project manager, if it doesn't severely affect the overall project objectives. Forcing, on the other hand, often proves to be a win-lose mode. Even though the project manager may win over a specific issue, effective working arrangements with the "forced" party may be jeopardized in future relationships. Nevertheless, some project managers find that forcing is the only viable mode in some situations. Confrontation or the problem-solving mode may actually encompass all conflict handling modes to some extent. A project manager, for example, in solving a conflict may use withdrawal, compromise, forcing, and smoothing to eventually get an effective resolution. The objective of confrontation, however, is to find a solution to the issue in question whereby all affected parties can live with the eventual outcome.

In summary, conflict is fundamental to complex task management. Not only is it important for project managers to be cognizant of the potential sources of conflict, but also to know when in the life cycle of a project they are most likely to occur. Such knowledge can help the project manager avoid the detrimental aspects of conflict and maximize its beneficial aspects. Conflict can be beneficial when disagreements result in the development of new information which can enhance the decision-making process. Finally, when conflicts do develop, the project manager needs to know the advantages and disadvantages of each resolution mode for conflict resolution effectiveness.[12]

Team Development Tasks	Personal Issues Facing Team Members	Examples of Team Building Actions for Each Development Phase
Phase I		
Climate Setting / Initiating What's our mission? How will we get started?	*Inclusion* How will I fit in? What will the other team members be like?	Clarifying the team's mission "Signing on" and including members Establishing a positive climate Building team identity
Phase II		
Goal Setting / Work Planning What specific goals need to be developed? What roles need to be performed? What team procedures do we need?	*Power & Conflict* Will my contribution count? How can I influence the team? How can I maintain my identity & also be a "team player"?	Identifying goals Assigning & negotiating roles Identifying needed team procedures Determining how the team will relate to the "external world," e.g. management, the sponsor, functional groups Balancing team and individual recognition Managing conflict
Phase III		
Implementing How do we keep up momentum & team morale? How do we maintain performance standards? How do we stay on track?	*Personal Performance Within a "Larger System"* What standards will I try to meet: mine? the team's? the organization's? How do I keep up my energy & interest? How do I relate to the team manager.	Assessing the impact of team norms on performance Assessing on-going performance Encouraging lagging performers Keeping team progress visible Reaffirming team-leader trust
Phase IV		
Evaluating Team Results / Following Up Did we meet our performance expectations? How do we sell team results?" What are our next steps?	*Dealing With Success / Failure and Transition* How has my contribution been evalutated? What have I learned? How satisfied am I? How do I let go of the team?	Reviewing team progress Sharing experiences and feelings Consolidating the learning Recognizing individual contributors Celebrating Providing closure

Figure 5.7 A framework for team development. (Source: Judith Mower, and David Wilomen, "A Framework for Developing High-Performing Technical Teams," *Engineering Team Management.* David I. Cleland and Harold Kerzner, Van Nostrand, Reinhold, New York, 1986, p. 300. Used by permission.)

5.4 A FRAMEWORK FOR PROJECT TEAM DEVELOPMENT

Mower and Wilemon have developed a useful framework for team development. The ideas that emanate from that framework are called *actions for team building,* and state:

> Team development is a process similar to individual development. Just as people mature through certain phases, teams go through noticeable phases. Teams mature in terms of task progress and in terms of interpersonal relations. At each phase of development certain problems can occur. Team managers can stave off such problems through team building. In this way team progress is not slowed.[13]

Mower and Wilemon's framework (Figure 5.7) describes four phases of task development and four phases of interpersonal relations development. In addition, typical examples of team-building actions used in each phase are briefly identified. This framework also applies, in micro-cosm, to planning and running team meetings.

In summary, the ability to create and lead a project team that is focused on well-understood objectives with a sound project plan, using appropriate operating procedures and integrative planning and control methods, lies at the heart of effective project management.

6

‹ ›

Building Commitment
in Project Teams

6.1 THE IMPORTANCE OF COMMITMENT
IN THE MATRIX

Since the project manager role usually must be carried out within a functional organization structure (for the total project at the owner or sponsor level, or pieces of the project at the contractor, subcontractor, supplier, or other contributing organization level), the resulting matrix creates a number of difficulties. These include confusion regarding responsibilities and authority, conflicts between functional and project goals and priorities, and related problems.[1]

One of the greatest challenges, or sources of difficulty, for the project manager in the matrix situation relates to *commitment:* First, how to get the commitments from the functional contributors that are necessary to achieve project objectives; and second, how to assure that these commitments, once made, are in fact fulfilled.[2] This is the human management part of the task—getting behavior to reflect project goals, priorities, and-interdependencies.

This chapter has been adapted from Rossi, Gerald R., and Russell D. Archibald, "Building Commitment in Project Teams," *Project Management Journal, XXIII,* 2, (June, 1992). The authors wish to thank Dr. Frank Wagner for his contributions to the initial development of some of the concepts presented in this chapter.

The project manager's job can be viewed as comprising two parts:

1. *Planning and Control.* Setting project goals and objectives; developing integrated plans, schedules and budgets; achieving the best allocation of available resources; authorizing and controlling the work; monitoring progress, replanning, identifying variances, taking or causing corrective actions.

2. *Leadership.* Influencing others and developing a sense of commitment in all persons contributing to or involved in the project.

Project plans and schedules are important prerequisites for gaining commitments on projects. Functional managers and project leaders have to know *what tasks* they are committing to and *when* the resources required for these tasks will be required, before they can make a serious personal commitment to the project manager. Progress must be monitored, plans and schedules revised, and any scope or schedule changes communicated in order to assure that the functional contributors honor their commitments. Any changes in past performance and future plans must be communicated to those whose commitments are affected by these changes. These topics are discussed extensively in the project management literature, but their importance with regard to commitment has not been widely recognized.

Two levels of commitment on projects can be identified: *organizational,* primarily indicated by the commitment of money, people, and other resources to a task or project; and *personal,* indicated by an individual's sense of dedication to assigned responsibilities, tasks or projects. Our discussion applies to both of these levels, but we will focus primarily on the personal level of commitment. (Personal aspects of commitment are also involved in organizational commitments, since these are made by persons acting for the organization rather than for themselves as individuals.)

This chapter provides project managers, project specialists, functional managers, and general managers with an explanation of the role of commitment in project management: its importance, the behaviors related to it, the major areas where commitments are required, and what a project manager can do to maintain the proper balance in managing commitments on projects.

6.2 LEADERSHIP AND COMMITMENT

While much has been written about leadership and its importance in management, little information has been published about the role of commitment in the leader's ability to lead, and its logical counterpart in followers.[3-6] Exacting commitment from oneself and from others are opposite sides of the same coin; together they provide the necessary energy to bring together the resources and motivation of a project team.

Before commitment can be developed in others, the project manager must provide vision, direction, and support to the project team members. *Providing vision* consists of defining and communicating direction, goals, and values in a way that is clear and understandable to all members of the project team.[7] This vision is usually communicated at the corporate level through the strategic planning process, policy manuals, and mission statements although it can just as easily be communicated through the "language of the (organization's) surrounding culture."[8] At the operational level, however, projects are the building blocks for implementing corporate vision. Creation and selection of projects to be funded, or choosing which requests for proposals for new projects to respond to, unambiguously communicate the vision that top managers have for the organization. At the project level, it is equally important to provide vision through clearly stated objectives and through communicating how the project supports the higher level objectives and strategies of the organization.

Developing commitment also requires giving individuals the *direction* and *support* needed to carry out their responsibilities. This second set of needs is satisfied through job descriptions, annual and quarterly objectives (MBOs), budgets, and so on, for the structured, functional organization. At the project level, individuals are given direction and support through the basic planning and control tools and procedures: Statements of project objectives and scope, the project/work breakdown structure, the project master plan and schedule, detailed task plans and schedules, integrated evaluation and control procedures, the task/responsibility matrix, and the like.

All of these topics are discussed in other chapters and, while they are necessary to successfully manage a project, they are not sufficient. There must also exist a strong *sense of commitment* on the part of those who will do the work and accomplish the goals. Our focus in this chapter is on the means for obtaining commitments and assuring that those commitments are fulfilled.

While commitment is critical to any endeavor, whether individual or group, little information is available about what leads to commitment, and even less about how to manage it. In the following sections, we will examine what is required to get the commitment of others and how the project manager can be a critical factor in managing that commitment.

6.3 UNDERSTANDING COMMITMENT

Most management research has focused on "organizational commitment"—identifying the determinants of organizational loyalty and their relationship to turnover of personnel. With a few exceptions, the relationship between commitment and organizational performance has essentially been ignored, especially with respect to how it applies within the culture of project management and matrix environments. We will examine what commitment is and how it can be directed to improve project management effectiveness.

In discussing the process through which an individual becomes committed to organizational beliefs, Sathe observes that "he or she internalizes them, that is, when the person comes to hold them as personal beliefs or values."[9] For our purposes, individual commitment is defined as consisting of two separate facets: (1) having a clear set of beliefs, values, and goals; and (2) behaving in a manner that is consistent with those beliefs and goals. Commitment is knowing where you want to go and being persistent in your efforts to get there. Project managers must be able to demonstrate a high level of commitment in their own behavior and to instill an equal degree of commitment in those with whom they work. Winograd and Flores describe an effective organization as one that has a strongly woven web or network of commitments that are made, observed, and honored, and that the manager's role is to create, coordinate, and enforce those commitments.[10] Members of a project team cannot be expected to have strong commitment to the project unless (1) the objectives are clear and understood, (2) a plan and schedule exists to which the team members can personally commit, and (3) the project manager displays strong commitment within their areas of responsibility.

Development of commitment takes place throughout the life cycle of a project. It can be thought of as a process of socialization and group development through which the project members develop and reinforce specific values and norms. This socialization can be accelerated through the use of "start-up workshops" to more quickly obtain the buy-in of project team members and clients.[11] (See Chapter 11.)

6.4   KEY BEHAVIORS FOR MANAGING COMMITMENT

Understanding the importance of commitment and what it means is not sufficient to elicit commitment from others. As with any other aspect of project leadership, managing commitment requires specific skills that individuals may or may not already have in varying degrees. Even those individuals with well-developed skills often have what might be termed "unconscious competence." They are successful but are not fully cognizant of what it is that makes them able to gain commitment from others. To maximize their effectiveness, project managers must become more "consciously competent" about how they can best manage commitment.

Two general types of behaviors, in combination, are needed to promote commitment—*supporting* behaviors and *innovating* behaviors. Supporting behaviors are those that lead to and build overall commitment. Innovating behaviors are those that create opportunities and the desire to exceed initial performance expectations and goals through improvements. Each of these two classes of behavior can be further divided into more specific skills (Figure 6.1). A balance among these behaviors is needed to obtain commitment from others.

Supporting Behaviors

Four key supporting behaviors build project management commitment:

1. Focusing on what is most important.
2. Demonstrating through example.
3. Rewarding contribution and results.
4. Managing disrespect.

Focus on what is most important. The first behavior, *focusing on the most critical factors,* relates to identifying those activities and goals that deserve the highest priority and allocating time and resources accordingly. This is an application of Pareto's law (the 80/20 rule) which states that in terms of return on investment of time, and effort, the most significant factors in a given group of factors normally constitute a relatively small proportion of the total. Usually, 80 percent of the results obtain from approximately 20 percent of the factors. While this law would seem to state the obvious, many managers do not effectively communicate and reinforce these differential priorities sufficiently through their own behavior. As we all are aware, it is easy for relatively unimportant factors to take up a disproportionate amount of our time or resources. For example, in one situation on a large engineering construction project, the project manager very much enjoyed getting out in the field and personally inspecting the job progress. At one point, it was obvious that the most critical need the project had was to devise a new plan to recover from the impact of an unexpected delay by a vendor in shipping a vital component. Instead of concentrating on this planning task, the project manager decided to go ahead with a previously planned, routine field trip to the job site. He was not focusing on the most critical factors by taking that action.

One of the project leader's key responsibilities is to make sure this does not happen—not only for himself, but also for all the project team members. As we will see in the next section, this requires more than just the

Supporting	Innovating
Focusing On What Is Most Important	Searching for Improvements
Demonstrating Through Example	Challenging Expectations
Rewarding Contributions and Results	Creating an Open Environment
Managing Disrespect	Encouraging Risk Taking

Figure 6.1   Key commitment behaviors.

ability to communicate priorities clearly to others. It requires managing one's own time and resources as well.

Demonstrate through example.   The second key behavior required of a project manager is *to be a visible example* of commitment. This means being a role model for the other members of the team. Albert Schweitzer once said: "Example is not the main thing in influencing others—it is the only thing." Although this may be an exaggeration, it speaks to the importance and impact of leading by example.

As discussed above, it is crucial to unambiguously communicate priorities to others. Often the most effective way to do this is not through what is said or written into planning documents; it is by letting your actions speak for you. Others will infer your real priorities more from your example than from what you say in your formal role as project leader. It is therefore important that what you do, how you do it, and where you spend your energy, reflect the priorities you want from the team. If you tell your team that it is important to plan and schedule the project using a particular microcomputer-based critical path method software package, and then never look at the schedules produced with this package, and make key decisions without reference to these results, it will be clear to the team members that planning and scheduling does not interest you, and they will conclude that it is not very important for them to plan and schedule their work in that way either.

For example, on a major project to develop and launch a new high-technology product, the project manager prided herself on her technical knowledge of the circuitry in the product. She accepted the corporate directive that required the use of specific project planning and scheduling procedures and supporting software, and encouraged her team members to also use these tools. When a major problem developed relating to the circuitry that threatened to cause a significant delay in project completion, she met with the project engineer and thrashed out a solution which she felt would minimize the delay. However, in doing so, she depended solely on her knowledge and experience, and never used the planning and scheduling system to check out the various alternatives on a "what if" basis. It became obvious to the other team members that she was not committed to the use of this system, and they in turn failed to use it to report their progress or rely on the system outputs to schedule their own work. Coordination and cooperation deteriorated, and when the project was finally completed, it was several months late.

Being an effective role model for project commitment is dependent in part on certain style factors. There are three distinct aspects of project managers' styles that make a difference in their ability to obtain commitment from others through demonstration. The first of these is demonstrating self-confidence. People are more willing to commit to someone who communicates a sense of confidence in their own abilities to success-fully manage a project through to completion. As a leader, showing doubt, worry, or fear can only undermine the confidence and commitment of others.

The second aspect, a corollary to the first, is maintaining an appropriate degree of humility. This does not mean self-deprecation. What it does mean is being willing to give credit to others and not being seen as egocentric and using a project solely for your own career advancement. Knowing when to ask for help in resolving a difficult problem is equally important to having a high degree of self-confidence. Balancing self-confidence and humility is not easy. It can be achieved, however, if the focus remains on the project goals and on helping others to be successful in their tasks, and if the project manager is aware of the limits on his or her own areas of personal competence.

The third behavior that contributes to providing a positive role model is encouraging and accepting constructive criticism from others. One of the keys to avoiding costly delays and other problems is maintaining open lines of upward communication. Members of the team have to feel that they can communicate their concerns about decisions, problems, or conflicts. This may be the greatest challenge facing any manager—accepting constructive criticism without becoming defensive.

The price of being in a leadership role is that the manager's own behavior is always under intense scrutiny, especially by those who report to him or her. Perhaps the worst thing any manager can do when leading others is to engage in the game of "do as I say, not as I do." One important reason for this is that it reflects directly on their integrity. In all organizations high integrity is strongly correlated to managerial effectiveness.

A final important outcome from providing others with a positive role model for their own behavior is this: when a project manager does not have the time or the ability to communicate, verbally or in writing, the subtleties of what he or she would like others to do (in an emergency, for example), team members can more easily make decisions consistent with those that would be made by the project leader.

Reward contributions and results.   The third supporting behavior is *rewarding contribution and fulfillment of commitments*. It is universally recognized that positive reinforcement is the most powerful motivator of people. Project management, by nature, often focuses on what is *not* on schedule. Unless a manager is careful, it is easy to emphasize the negative and what people have not accomplished rather than their positive contributions. In a matrix environment, several types of contribution deserve reward and recognition. Results are paramount. Achieving major milestones on schedule and under budget are the best indications of high performance—if the quality of results is also there.

In addition, however, people need to be recognized for other contributions. One of these is effort. Although effort is not as good as results, it is an indicator of commitment. The complex nature of some projects sometimes makes the achievement of results difficult, if not impossible. Project team members who put out a concerted effort against difficult odds may deserve as much recognition as those who accomplish easier goals. This is especially true during the course of long-term multiyear projects.

People deserve to be rewarded for fostering teamwork and collaboration. In a project environment, the climate must support cooperation among individuals with different values, skills, and perspectives. It takes a conscious effort on the part of everyone involved to keep individual conflicts at a minimum and to focus on the team's goals. While this is one of the project manager's primary responsibilities, he or she cannot be expected to carry the entire team-building burden. Those who help foster a climate of cooperation should be recognized and rewarded for that contribution.

Manage disrespect.   The final key supporting behavior is *managing disrespect.* By disrespect we mean any comments or actions that communicate a negative attitude or feeling about something. A common behavior of many people is to make derogatory comments or jokes about something or someone connected with a project. The typical objects of derision are coworkers, their functional managers, the project manager, the project sponsor, the organization, their own tasks, or the client. While this type of behavior is found in all kinds of organizations, it can be especially prevalent in a matrix. The natural conflict of interests and differing perspectives among team members makes such behavior common. The project manager's responsibility is to manage this behavior by eliminating it if possible, and minimizing it everywhere else.

Although many disrespectful comments are intended to be humorous, this type of humor has a dark side that can quickly undermine commitment. As a leader, the project manager must neither engage in disrespectful comments nor allow others to do so. This is not to say that there is never anything to criticize or that it is wrong to identify areas where problems or conflicts exist. What it does mean is that there is no excuse for derogatory and insulting comments. Making such negative comments only undermines the manager's own credibility and integrity and reduces the level of commitment in others. Dale Carnegie summed this up nicely with his three Cs: "Never condemn, criticize, or complain!"

Innovating Behaviors

Commitment is helpful only up to a certain point. There are times when commitment can be dysfunctional. That time arrives when an individual's inflexibility prevents him or her from recognizing or searching for a better way to accomplish goals. Any commitment must be balanced with a willingness to change when necessary. In project situations, trade-offs are often required between project scope, quality, performance, cost, and schedule, especially when unexpected and unplanned events occur. A team member who remains totally committed to the original plans, regardless of changing circumstances, is not effective. Because of the need to frequently respond to changes, the project manager and project team members need to understand the higher level vision, objectives, and strategies which the project is intended to support, in order to respond most effectively to those changes. This can also be described as management innovation.

We have identified four behaviors to be important in managing commitment to innovation. They are:

1. Searching for improvements.
2. Challenging expectations.
3. Creating an open environment.
4. Encouraging risk-taking.

Searching for improvements.   The first behavior, *looking for a better way,* refers to the search for improvements in design, performance, costs, and schedules. One of the keys to successful project management is good planning. Planning, however, should not mean inflexibility. It means using plans as guides to decisions and as yardsticks for measuring performance. These purposes do not preclude making improvements in the way things will be accomplished, or the schedules for accomplishing them. Engineers are widely accused of wanting to continually improve a design, or a product, thereby increasing the project scope and causing delays and cost overruns. By improving through looking for a better way, we do not mean to encourage or condone such unwanted results. A design improvement is a "better way" if it produces a product with the same specified performance (or better) at less cost or within the same or shorter project schedule. We need to encourage management innovation—to find better ways to plan, integrate, perform the work, and otherwise get the job done—as much as we need to innovate in the product or engineering sense.

There are two major sources of new ideas for a project. One is from members of the project team; the other is from outside sources (either within the same organization or from other organizations). Some organizations use a team of "old hands" to look over the shoulder of the project manager and pass on their experiences to help find better ways of doing the job. Most innovations result from new applications of existing ideas, combinations of currently available technology, or incremental improvements over previously used methods. These sources of fresh ideas are less costly and can be more quickly applied than solutions that must be developed from a clean sheet of paper. In many projects, such outside ideas fall prey to the "not-invented-here" syndrome. Teams will reject ideas developed by others and choose instead to expend time and resources to develop a new solution of their own. Effective project managers encourage their team members to accept new ideas from any source. The criterion for judging an innovation must be its value added, not where it comes from.

Challenge expectations.   The second necessary improving behavior is *challenging expectations.* Individuals and teams often limit their own performance as a result of their expectations of what can be reasonably achieved.[12]

For example, a football team may set as its goal competing in the World Cup matches, or—in American football—"getting into the Super Bowl." The more appropriate goal should be to *win* the World Cup or *win* the Super Bowl. One American football team examined why it had lost in the Super Bowl, and discovered that the team had set as its goal merely to get into that final game; they realized, after their loss, that they had not set the proper goal for themselves: To win that final game.

If these self-limiting expectations by project teams, just as by sports teams, can be identified and challenged, performance will often exceed initial estimates. Primary responsibility for doing this rests in the project leader. He or she must recognize where and how the project team is constraining its creative potential, and eliminate those constraints.

Create an open environment.   In addition to searching for improvements and challenging expectations, a project leader must be *open to change* and create a climate which encourages others to be similarly open. Project environments, more than most, are resistant to changes in plans. The complexity and tightly linked interdependencies of large projects creates an aversion to any change, even when it would improve results. Deviations from plans are seen as costly and fraught with risk. Being unwilling to make changes when they are appropriate, however, can be equally dangerous. Alfred North Whitehead expressed the essence of the quandary when he said, "The art of progress is to preserve order amid change and to preserve change amid order."

The Tarbella Dam on the famous Indus River, near Peshawar and the historic Khyber Pass in northwest Pakistan, is the largest earth-filled dam in the world, with a volume of 175 million cubic yards. (The largest in the United States is Fort Peck Dam, with 125 million cubic feet.) It feeds a major hydroelectric generating facility of great importance to the economy of Pakistan, and the water downstream of this facility irrigates a large part of the country. After the dam itself had been completed and the reservoir was being filled, and after the main power intake tunnel (reinforced concrete 30 feet in diameter, over a mile long) had been completed, the project manager, a Pakistani engineer for the Pakistan Water and Power Development Authority, took an innovative, risky step that paid very big dividends. In reviewing the data on the volume of water that was flowing into the dam, he noticed that it exceeded the original forecast significantly. He determined that two additional turbine-generators could be driven with this additional volume, with an increase in electrical generating capacity of 25 percent over the original design. The problem was that the main intake tunnel had already been put in place. Through innovative engineering design, a bifurcation or dividing device was placed in this tunnel that enabled diversion of the required amount of water to the additional turbines. This major change in project scope at a late stage of the project caused delay in the completion time and added to the cost, but it proved to be a wise investment. The project manager risked severe criticism, in a culture that does not foster individual initiative, in pushing for this innovative change, but he gave his country a huge amount of additional electrical energy over a period of many years by doing so. The energy in the larger flow of water would have gone untapped into the Arabian Sea if he had not taken this risky initiative.

Encourage risk-taking.   Finally, the project leader must be able to *encourage and support taking reasonable risks.* Attempts at improving any aspect of a project always involves some degree of personal risk, as well as risk to the project itself. Throughout the history of organized group activity, individuals have usually been admonished or otherwise criticized for taking new and innovative paths, whether or not they have been officially encouraged to do so. Often individuals will be punished for taking risks even when it results in positive outcomes. (There is seldom any doubt about the reaction of higher management when taking a risk results in failure.) The alternative for receiving a medal for heroism in military service can be a court martial and disgrace.

As a result, through experience, people learn to avoid taking risks. When making changes they choose the path that minimizes their own exposure.[13] Unfortunately, actions that minimize such downside risks will often also minimize opportunities for creativity, innovation, and improvement.

Project leaders want members of their team to take risks, but they should be "reasonable" risks. This means, to be acceptable, risks must have three fundamental characteristics. First, expected benefits must clearly outweigh expected costs. Second, risks should only be taken in areas where there is sufficient slack in both time and resources to be able to recover if things don't work out as planned. Third, the decision to take risks, and the degree of risk involved should depend on the confidence of the project leader in those who are proposing the risk, based at least in part on their prior record of successes.

The ultimate test of how effectively a project manager is able to encourage and support risk is measured by the way he or she deals with failure. If you want risk-taking behaviors, you must also be willing to accept a certain (but limited) number of failures. A project leader's commitment to improvement, and the risk-taking behavior needed to achieve those improvements, will not be judged by the way he or she manages successes. It will be judged by how they manage failures. Failures must be managed in a way that will encourage continued reasonable risk-taking in the future.

6.5 WHERE TO APPLY COMMITMENT BUILDING BEHAVIORS

Up to this point we have only discussed *how* a project leader must act—what behaviors he or she must exhibit day in and day out to ensure that the project team achieves results on schedule and under budget. There is a second important part to the equation, *where* these behaviors should be applied—the targets of the eight commitment behaviors discussed above.

There are six key areas in which project leaders should be practicing commitment building behaviors. Each of these areas must be continually *supported* and when necessary, *improved*. As shown in Figure 6.2, they are:

1. Organizational values.
2. The organization and higher management.
3. The project clients and sponsors.
4. The project goals and tasks.
5. The project team and its members.
6. Themselves as project managers.

Commitment to values.   Key organizational values must be communicated and enforced. This may be the most important area in which the project leader should be a role model. The pressures of project management can easily lead otherwise ethical individuals to compromise values to get results. There are many recent examples, especially in the defense industry, of what can happen without constant attention to maintaining and improving adherence to basic values. What is the result on the team member's commitment when the project manager asks them to charge their time incorrectly to another project, in order to minimize a cost overrun on his or her project?

Commitment to the organization and its management.   The second commitment area is the organization and higher management. The ulti-

Key Areas of
Commitment

Figure 6.2 Key areas of project commitment.

mate purpose of any project is to enhance the long-term success of the organization and to contribute to the goals and strategies of top management. In this area, the project leader must maintain commitment to the organization of which he or she is a part. Without building this commitment, a project leader cannot expect to receive the support from others necessary to achieve long-term results. The earlier comments regarding managing disrespect apply here as well.

Commitment to the project clients and sponsors. While it may seem self-evident that a project leader must display commitment to the project's client and sponsor(s), it is not always evident in his or her behavior and in how he or she manages the team. An obvious indicator is the way they manage any disrespect (as described earlier) for the client. When problems occur or plans must be changed to accommodate changes to the needs of the client, it is easy to react by putting the blame on, and making (or allowing others to make) derogatory comments about the client. This can be viewed by others, inside and outside the project team, as a lack of commitment and respect. Such a perception can undermine the team's long-term commitment to the client.

Commitment to project tasks. Along with everything else, there must also be a strong commitment to project milestones and tasks. These represent the short-term day-to-day aspects of the project. Achieving high performance in a project environment demands attention to details. Improving overall results is the product of maximizing results, and seeking improvement, in each component of the total plan.

In one case, the project manager of a medium-sized information system project had established, with her team, a reasonable number of key milestones throughout the project plan and master schedule. Completion of several of the early milestones was delayed, due primarily to lack of attention, or commitment, since the work at that stage was not extremely demanding. However, the team members convinced the project manager that they could recover these delays on future, much more demanding and complex work. They even reflected the recovery in their detailed plans. In the end, the project completion was delayed by almost the exact amount of the early delays. If the team had held to their original commitments on the early milestones, the project would have been completed on schedule. Instead the project was late, creating significant cost overruns for both the project organization and the client.

Commitment to the project team members.   Although project managers generally have only a matrix relationship with members of their team, they can still be an important influence on the development of both individual talent and team effectiveness. Team members want project leaders who believe in their team and who are concerned with the success and development of individuals as well as the realization of the total project. To successfully elicit a high level of commitment from others, a manager must display a strong sense of commitment to the members of his or her team. Project managers frequently have the opportunity to "run interference" with higher managers for the team, protecting them from the frequently unreasonable demands imposed from on high. Team members will recognize this commitment to the team, and it will reinforce their own sense of commitment as a result.

Commitment to themselves as project leaders.   Finally, an effective project leader also demonstrates a strong commitment to themselves. This does not mean self-interest or egoism. It does mean exhibiting the right balance between self-confidence and a willingness to listen to advice and constructive criticism from others. This is perhaps the project leader's greatest challenge—remaining open to feedback from others about how he or she can improve.

6.6   BALANCING COMMITMENTS

Plato described the road to excellence as being the "golden mean," a path defined by a flexible balance among conflicting and sometimes incompatible demands. The same can be said for achieving excellence in project leadership. First is the need to maintain a continuously shifting balance among the conflicting demands imposed by the values of the organization,

Figure 6.3   Balancing goals, values and needs.

the demands of the clients and sponsors, the requirements of the tasks, the constraints of the schedule, the dynamics of the team that must perform the tasks, the responsibilities to the overall organization, and, finally, by the personal goals and values of the project leader (Figures 6.3 and 6.4). Maintaining equilibrium among each of these conflicts is itself exceedingly difficult. Keeping them all in balance simultaneously is one of the major challenges of the project manager and requires constant attention and adjustment. It is in this area that experience and political awareness are especially important.

A second challenge is to determine the appropriate balance between supporting and improving behaviors. How much effort should be placed on achieving the plan as currently defined, and how much on striving to improve on those original goals? How much on rewarding goal achievement versus taking risks on new ideas? How to manage disrespect while encouraging others to challenge expectations when it is necessary? There is no simple answer except to say that the project leader's behavior will do more to set the tone and direction in these areas than will any written statements of guidelines.

Figure 6.4   Balancing behaviors.

In attempting to maintain an appropriate balance among all these factors, values play a critical role. Conflict and its effects can decrease an individual's, or a project team's, sense of commitment. When such conflicts occur a clear set of organizational values, applied consistently, can help each individual team member find the right balance in carrying out their own responsibilities. The project leader must through his or her behavior, simultaneously be both the messenger of and the role model for those values. Without an attention to balance, a project organization can fail to serve any of its stakeholders.

6.7 IMPLEMENTATION

Translating these concepts into operational behaviors requires more than just raising the awareness of project managers to their importance. It requires a systematic process to assure that every project leader (and if possible every team member) has the knowledge and skills necessary to implement the approach we have described. Accomplishing this requires the following five steps.

1. Assessing each project leader's capabilities related to supporting and innovating behaviors in each of the key areas discussed (values, organization, client, project goals and tasks, people on the team, self).
2. Identifying indicators of the key behaviors necessary for effective project implementation.
3. Initiating a feedback and follow-up process for providing individual and group feedback on performance with respect to supporting and innovating behaviors within key commitment areas and measuring their progress over time.
4. Providing management development opportunities designed to enhance both the organization's and project leaders' capabilities to manage commitment.
5. Developing necessary evaluation mechanisms and follow-up to ensure continuing support of project commitments.

The above list is based on a general model of how project organizations work. Each step is part of a process (see Figure 6.5). The first stage involves a preliminary assessment of capabilities with respect to managing change and implementation. The commitment framework provides a starting point for identifying critical strengths and weaknesses in the project organization's capabilities that influence its ability to achieve project goals.

The outcomes of this first step provide data for the development of an organizational profile which permits comparison of capabilities directly to project requirements. This includes identifying behavioral changes needed

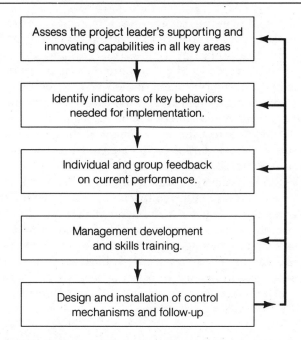

Figure 6.5   Implementation process for behavioral changes.

to support the achievement of specific short-term milestones and long-range goals. Also important at this stage is consideration of the organization culture and how it will impact communication, teamwork, and cooperation.

The third step consists of designing a feedback mechanism which reflects key project implementation issues. This is accomplished through a multimethod approach using interviews of key managers at all levels of the relevant organizational units, collecting data from existing sources (internal and external), analyzing project requirements for the proposed plan, and reviewing results of the preliminary survey and profile developed above. This phase focuses data gathering around specifics that will affect the organization's and the team's ability to carry out project strategies.

The fourth step involves communicating project strategies (if necessary), providing individual and group feedback, building skills, and reinforcing organizational strengths. This is usually accomplished through surveys and implementation workshops for key project, client, and sponsor personnel. These workshops are designed around the general commitment framework described above but focus specific attention on factors that are either (a) central to project success or (b) identified as an area of weakness in the earlier "capabilities assessment" phase. As part of this process (internal or external), facilitators/consultants conduct both individual planning sessions as well as a group-based upward communication process.

In the individual planning sessions, each manager uses their own individual feedback reports to analyze their strengths and areas for improvement with respect to project goals and strategies and develop a personal action plan. This plan specifies how they intend to improve their capability to contribute to the accomplishment of their responsibilities in the project. The upward communication process provides each team with an opportunity to give top management feedback on existing or potential problems and issues that might affect the achievement of the project goals. In addition, composite feedback data provides an assessment of the overall project organization's current capabilities.

The final step is to develop organization-level support plans based on the results of the survey data, feedback from key members of management, and from the upward communication process. Possible areas for action here would be such things as the performance review system, resource allocation plans, management and skill development programs, and adjustments to the matrix structure. It is hard to predict what kinds of issues and problems will surface at this stage. Each organization's situation will have its own unique set of problems and constraints that will in turn determine the optimal set of solutions.

6.8 SUMMARY AND CONCLUSIONS

To be successful, project leaders must develop commitment in an environment with a variety of conflicting demands and factors. That commitment is demonstrated by a balanced combination of what we have described as *supporting* and *innovating* behaviors. The keys to obtaining commitment from others are simple in concept but require constant attention and reinforcement. They are:

1. Providing others with a positive role model for all aspects of commitment to others (that is, leading by example).
2. Encouraging continuous feedback from everyone on performance, progress, and opportunities for improvement.
3. Giving adequate attention to both supporting current plans and striving for improvements.
4. Maintaining a balance among priorities.

Project managers who gain the commitment of their project team members, and actively manage that commitment throughout the planning and execution of their projects, will have a higher probability of achieving their project objectives.

7

« »

A Strategy for Overcoming Barriers to Effective Project Management

Introducing integrated project management practices and the related formalization of the project management function usually requires significant adjustments in attitudes, understanding, responsibilities, methods, and reporting relationships throughout the involved organizations. These changes affect the parent organization and all organizations represented on the project team.

Cultural and other factors—within the project environment, the involved organizations, the industry, the geographic region, and the involved nations—create barriers to these required changes. These barriers can require substantial effort to overcome or mitigate and, if they are not overcome, they will reduce the effectiveness of the project management efforts. This chapter outlines a five-phase strategy to implement changes required

This chapter has been adapted from Archibald, Russell D., "Overcoming Cultural Barriers to Effective Management by Projects," *Proceedings of the 10th INTERNET World Congress on Project Management,* INTERNET International Project Management Association, Manz Verlag, Vienna, 1990. Grateful acknowledgment is given to Dr. Gerard Rossy for his comments and suggestions.

127

for effective project management and to manage the barriers that will be encountered: (1) Identify and understand the barriers anticipated in regard to a proposed change, (2) create awareness of the need for change and identify and harness the motivating forces that will help to overcome the barriers, (3) educate and train all affected people using the knowledge gained in the first two steps, (4) define "change projects" to implement new project management practices and use good project management practices to plan and execute them, and (5) modify and evolve the project management practices and/or the manner of their implementation to accommodate the current or anticipated cultural and other barriers.

7.1 KEY CHANGES REQUIRED

The basic project management practices used to identify potential barriers to change are those summarized in Chapter 1 and described in some detail throughout this book. The task of implementing practices based on these concepts requires changes in a number of areas, regardless of region, country, industry, or specific organization. Some organizations have made many of these changes, but few in any culture have totally adapted as yet to these concepts. Some of the primary areas of change that generate barriers and require understanding and acceptance are:

1. Integrative roles below the general manager.
2. Shared responsibilities for projects.
3. Direction from two bosses: functional and project.
4. Integrative, predictive planning and control.
5. Computer-supported management information systems.
6. Project objectives over department objectives.
7. Working, and being rewarded, as a team rather than as individuals.
8. Temporary assignments on projects, rather than permanent niches in the bureaucracy.

In order to overcome barriers to change, each organization needs first to *identify and prioritize* the key changes, such as those listed above (the list is by no means complete), that are required to progress toward full management by projects. Then the barriers to each of these changes can be identified so that strategies for mitigating them can be developed and executed.

7.2 IDENTIFYING BARRIERS TO THESE CHANGES

Integrative Roles below the General Manager

In the functional organization, the president, managing director, or chief executive officer plays the key integrative role at the corporate level, and general managers (or equivalent titles) play this role at lower levels, such as operating divisions. Project managers are assigned within the organization as integrators for their projects, and functional project leaders integrate the work on a specific project within their departments.

Barriers.   These integrative assignments of the project manager and functional project leaders create barriers that stem from the bureaucratic structure of many organizations. Since the general (or higher) manager traditionally has been the primary integrator, a subconscious feeling is that these new integrators are trying to act like, or take over the authority of, the general manager. Because the project manager is obviously not the general manager, people will often resent and resist this role, unless they fully understand it. The functional project leader likewise is not the overall functional manager, but acts to integrate the work for the project within the function. The more authoritative the organizational and social culture, the more difficult this barrier will be to overcome. These barriers can be made worse when the project manager thinks he or she must act like the general manager and tries to assert full authority over all project team members and their responsibilities.

Shared Responsibilities for Projects

For effective management by projects, both functional and project managers and functional project leaders must learn how their responsibilities are properly shared on projects. In oversimplified terms, the project manager is responsible for *what* (project scope) and *when* (project schedule), while the functional managers/leaders are responsible for *who* does the work and *how* the work is performed. *How much* (the project budget) is the responsibility of the project manager, but is usually based on the functional estimates.

Barriers.   The culture of traditional functional organizations, with each manager totally responsible for the what, when, how, and how much, is a source of barriers due to lack of understanding and acceptance of this sharing of responsibilities that is required for effective management by projects. The functional managers feel they have lost power and prestige, and do not like the feeling or the fact. In national cultures that place a heavy emphasis on power, control, and hierarchical position, this barrier is very strong. Such shared responsibilities may be perceived as a sign of weakness for both parties.

Direction from Two Bosses: Functional and Project

Project team members usually must take direction from a functional boss (how to do the work) and from a project boss (what to do—scope of task—and when to do it to meet the project schedule).

Barriers.   In traditional corporate or national cultures where people have learned that every person should have only one boss, this attitude produces another barrier to effective project management. Additionally, some team members may play one boss off against another to cover up problems, aggravating the obstacles mentioned earlier.

Integrative, Predictive Planning and Control

Project planning, scheduling, and budgeting tools and methods are designed to integrate information and predict outcomes if the project plans are followed.

Barriers.   People in bureaucratic organizations have learned that "information is power," and therefore have developed attitudes and habits which causes them to "play it close to the vest," and hold back information about their work. Some corporate and national cultures do not encourage the sharing of information. Information access may be linked to position in the hierarchy rather than the person's need to know it. These attitudes, habits, and customs cause resistance to the new methods that are designed to integrate everyone's plans for the various tasks on a project, openly communicate the status of everyone's work, and reveal problems or conflicts in a predictive way, so that they can be resolved or avoided. These obstacles are added to the aversion that many technical people have to planning in any form.

Computer-Supported Information Systems for Management Purposes

Computer-supported information systems, especially those using microcomputers, are becoming more useful and necessary for effective management by projects.

Barriers.   Older managers, and managers from cultures who view a keyboard as something only a low level clerk would ever touch, have difficulty understanding how such systems are useful for management purposes. Other managers view computer systems as only for accountants or engineers, not managers. Sometimes the "computer experts" promote and exploit the mystique of computer systems, and multiply these barriers. The traditional barriers between technology and management must be broken down.

Project Objectives over Department Objectives

Most of us, in all cultures, develop a sense of loyalty and commitment to our own departments or other organizational units.

Barriers. For effective management by projects, we need to be able to recognize the importance of the project objectives, and even place these above our departmental and sometimes our personal objectives. The inability or unwillingness to develop a loyalty to the project, and become committed to its objectives, can be a significant barrier to its success, and to the teamwork required for effective project management. Some cultures base personal rewards almost solely on achieving departmental objectives and satisfying the functional manager, with no regard for the individual's performance on a project. Cultures that reward loyalty and status quo over performance and professionalism can be especially affected here.

Working, and Being Rewarded, as a Team Rather Than as Individuals

Successful projects require all project contributors to work closely together as a team. However, most project team members have been successful in the past, and have been rewarded, for working as individuals.

Barriers. This need to change behavior and work habits toward multidisciplinary teamwork exposes barriers that prevent many individuals from wanting to, or being able to, work as members of a well-oiled team. In national cultures that have fostered and perhaps even glorified individual work, these barriers can be formidable. At one extreme, team solidarity can be a barrier to good performance when team members protect each other to the detriment of the overall project performance.

Temporary Assignments on Projects

Everyone seeks the security of a long-lasting, stable, predictable organization, and many of us find it less than desirable to be in a temporary, personally risky, less predictable project situation where the end is not only in sight, but great pressures are exerted to complete the project as soon as possible.

Barriers. After observing what can happen to the people assigned to a project that is abruptly canceled, or fails, or even when a project is successfully completed and the team members are rewarded with demotions or termination of their employment, people will not be motivated to follow in their footsteps. These barriers, as well as many of those described earlier, are primarily psychological, and they can be very powerful deterrents to accepting and using effective project management practices.

In addition to these specific project management-derived barriers, cross-cultural lack of understanding or long-standing animosities bring with them additional barriers. These can be found in joint-venture projects bringing together two corporate cultures in the same country, or projects involving two industries, or multinational projects involving two or more nationalities and languages. A number of additional cultural factors that create barriers to effective project management can no doubt be identified.

7.3 FORCES HELPING TO OVERCOME THE BARRIERS

A number of forces will usually be present in a given situation which, if harnessed, can assist in overcoming the barriers. Pressures from the organization's competitors, failures which show that the existing methods are not adequate, the interchange of people with different experiences from other organizations, the desire within many people to improve and do a better job, the need to prepare the organization for the future, dissatisfaction with the present situation—all of these plus others are sources of energy that can be harnessed.

Of critical importance is the need to create an *awareness of the need for the change* in question. This awareness is often the result of a project disaster, either within the company or the industry, that brings home to all concerned that products and markets are changing, environments are changing, competition is becoming global, and if we want our organization to survive, it is necessary for us to change also. Creating the awareness of need for a given change requires demonstration, education, and communication. Once the awareness of need exists, the barriers will be less formidable, and can be surmounted or mitigated more easily. In laying out a strategy for dealing with change, priorities must be set, considering both the importance of each factor, and the feasibility of making the change.

7.4   EDUCATION AND TRAINING

Having identified the expected barriers, and having gained an understanding of the forces that are present that will assist in making a change, one can design an education and training program to support the implementation of the change. The nature of project management, which involves discrete projects, allows the use of selected projects as the education and training vehicles. Such projects can be thought of as prototypes from the management viewpoint, testing the new practice or method, and demonstrating to the project team and the entire organization what the role of the project manager really is and how the changes being introduced in the project planning and control system really work.

Team planning workshops, described in Chapter 11, are being recognized more widely as a very effective method of (1) introducing new project management practices; (2) educating, training and developing the project teams; and (3) overcoming the barriers that may block the total effectiveness of the project management approach. One group of experienced professionals recommended four types of workshops for cross-cultural projects: Workshops at both the management and project team levels, and—for each level—two workshops, one dealing with the cultural aspects and factors, and the second dealing with the project objectives, scope, content, and plans.

It is not effective simply to announce the appointment of a project manager or the acquisition of a new planning and control system, for example. A well-designed education and training program is required, anticipating the barriers to be expected, and capitalizing on the motivating factors which will assist in the acceptance of the change. Middle-level managers generally resist the introduction of project management practices most strongly, so they must be included in the education and training program.

7.5   TAKING APPROPRIATE ACTIONS TO IMPLEMENT THE CHANGE

Either in conjunction with or following the proper education and training, management actions are required to introduce and put into practice the project management concepts described in this book. Senior managers themselves must learn about, understand, and support the new practices, and demonstrate through their communications and behavior that these are important to the future of the company. Senior managers must recognize that they are role models in the organization, and that their actions and attitudes speak far louder than their words. The introduction of new integrative roles, new planning and control methods or tools, and new ways of working together as teams, *must be viewed as projects in themselves.* We can "let the medium be the message" in this regard. These change projects must be carefully planned and must include supporting education and training efforts designed around the anticipated barriers. Recognizing these as "management research and development projects" can be a useful way of positioning them and gaining greater acceptance and support throughout the organization.

7.6 MODIFYING AND EVOLVING PROJECT MANAGEMENT PRACTICES

In all organizations that manage projects, their practices have evolved over a period of time. It is not possible or even desirable to attempt to leap from no formal project management to the full-blown ultimate that can be envisioned, or that some organizations have actually achieved over a number of years. Many of the cultural and other barriers are formidable and will not be overcome in a short time. In each of the three basic concept areas (integrative roles, planning and control, project teams), it will be necessary to introduce changes on a step-by-step basis. Compromises will often be required between the ideal and what can be made to work this year. Experience will be needed in absorbing one level of change before the organization is ready for the next. For example, an organization may elect to start out with assignment of a project coordinator, with a more limited integrative role, rather than a project manager, due to cultural barriers involved in the "manager" title, or in acceptance of the full integrative role. In such cases, the manager to whom the project coordinator reports will probably carry out the role of the project manager, in addition to his or her normal duties.

7.7 SUMMARY

Success in overcoming the cultural barriers to effective management by projects can be enhanced by using the five-phase strategy described here:

1. Define the changes required and identify the anticipated barriers.
2. Create an awareness of the need for change, and identify and harness the motivating forces that will help to overcome the barriers.
3. Educate and train all affected people using the knowledge gained in the previous steps.

4. Define "Change Projects" to implement new project management practices, and use good project management practices to plan and execute them.

5. Modify and evolve the project management practices to accommodate the barriers.

Project management is the management of change. Improving project management capabilities requires change. Therefore, implementing or improving project management itself requires the use of effective project management practices, and must be viewed from a long-term perspective. There is no one best answer that fits all situations. The concepts of project management must be tailored to the situation and culture, including the cultural mix of the project team.

8

« »

Multiproject Management

I t is rare to find a project that exists by itself without interaction with
other projects. The multiproject environment commonly found in
many large organizations imposes complications at the project manager
level, and at each level of the functional organization. The basic problems
result from competition between and within projects for resources and
management attention.

Since no known organization possesses completely unlimited resources,
each project manager faces one or both of the following situations:

- The project is supported by functional departments that are also
  supporting other projects, and hence there is competition for available resources.

- He or she is project manager for more than one project, each of which
  will similarly be competing for needed resources.

At appropriate levels in the organization, the multiproject requirements and priorities must be brought together to assure that all projects
are completed so that the maximum benefits are obtained for the entire
organization.

The need for some means of integrating these multiproject requirements
is frequently not fully recognized. The result is that a number of projects
are individually managed from different parts of an organization, with a
great deal of effort being exerted by the various project managers to win

136

every competition with the other projects when available skills, or other resources needed by the projects, are not sufficient to satisfy them all. A very strong project manager may thus be successful on one project, but the company may suffer severely from the delays produced in other projects.

This chapter discusses the objectives of multiproject management, related priority and resource management questions, and indicates how the tools described in Part II can assist with these management needs.

8.1 OBJECTIVES

The higher order objectives of multiproject management, in comparison to managing a single project, include:

- Completion of all projects to best achieve the overall goals of the organization.
- Determination of both long-term and short-range priorities between projects to enable appropriate decisions regarding allocation of limited resources.
- Acquiring and maintaining an adequate supply of resources to support all projects, including people, facilities, material, and capital; but at the same time assuring that these resources are gainfully and efficiently employed in approved, productive work required to complete the projects.
- Integrating these multiproject requirements with other ongoing activities not related to projects as such (production of off-the-shelf products, etc.).
- Developing organizational patterns and management systems to satisfy the ever-changing project needs on one hand, and to provide, on the other hand, organizational stability, professional development, and administrative efficiency for persons managing and supporting various projects.

8.2 MULTIPLE MAJOR PROJECTS VERSUS MULTIPLE SMALL PROJECTS

This book is directed primarily to situations involving a small number of major projects as they are defined in Chapter 2. However, of equal importance are the project management needs in the situation where a large number of relatively small projects exist. This is typical of many high technology companies that design, manufacture, and install complex products or systems.

To illustrate this point, consider the situation where one or more contracts with customers cover a large number of specific central office telephone exchanges or central telephone switches (PBX's) and systems on a user's premises. Each of these is in reality a project, requiring engineering work, manufacturing, installation, and test. These overlapping projects flow through marketing, engineering, one or more factories, and only emerge as clearly distinct projects during installation, where typically the installation department manages a number of geographically separate sites. In such situations, it is not possible or even desirable to appoint a project manager for each PBX with responsibility through engineering, manufacturing, and installation. In some cases a project manager is appointed for a group of related exchanges, or for a very large, complex PBX and system. The general practice was to appoint a supervisor for each exchange only during the installation phase, reporting to the installation department.

	Multiple Major Projects	Multiple Small Projects
Project manager role	Assigned to a manager who does not have a functional responsibility.	Retained within the line organization with integrated staff assistance in planning and coordination. Possibly several projects are assigned to a project manager or project coordinator.
Project team	Key team members may report to the project manager or may only be physically located together or may stay in their functional organizations.	Project work always assigned to functional departments; in field phase, full-time team is assigned to each project.
Integrated planning and control	Each project planned and controlled on an intergrated basis; multi-project conflicts resolved above project manager level. Different systems often used on different projects.	All projects must be planned and controlled on an integrated basis within one multiproject operations planning and control system; conflicts resolved by multi project and/or functional managers.
	Schedule control on each project most important; resources usually secondary.	Best use of limited resources usually most important; resource, constrained scheduling required. Priority control is important.

Figure 8.1   Key differences between multiple major and small projects.

Even though project managers may not be appointed to every project in this situation, management of the projects on an integrated basis is still of vital importance to meeting the contract cut-over dates. Since a typical telephone switching equipment division is predominantly devoted to the execution of these somewhat repetitive (but never identical) projects, the division general manager retains the overall project management responsibilities. This places even greater emphasis on the need for organization, methods, and systems that enable truly integrated planning and control of all projects through all their life-cycle phases.

Centralized planning and control offices have been established in some companies to provide the needed marketing-engineering-manufacturing-installation planning and master scheduling where this multiple small-project situation exists. Coordination of these plans and master schedules, and appropriate follow-up to assure compliance, is handled by a planning manager reporting to the division general manager. These offices are useful in training and developing individuals with the special skills required for project management support personnel on major projects.

Figure 8.1 summarizes the key differences between these two commonly encountered situations. Further discussion of the organizational aspects of the concept of an operations planning and control function is presented in Section 8.6 of this chapter.

8.3 PROJECT PRIORITIES

In a situation where several projects are competing for limited resources, the need for a method of determining and communicating the relative priority of each project is easily recognized. However, developing a method for satisfying that need has proved to be a difficult job.

Effective planning of each project and forecasting of its needs will assist in prediction of potential conflicts with other projects, if all are similarly planned. Decisions can then be made whether to supply additional resources or to delay one or more projects. Some measure of project priority is needed to make such decisions. The *prediction* of such potential conflicts is the only way to enable effective management action which can avoid the actual conflict in a crisis atmosphere.

Unforeseen problems or needs can cause short-term conflicts even in the most thoroughly planned projects. When these occur, the person allocating a limited resource to one or the other project must have accurate knowledge of their current relative priorities.

Few organizations have an established method for handling this question. In the absence of such a method, numerous decisions are made daily at the first line management and supervisory levels regarding relative project priorities. It is conceivable that contradictory decisions are made on different parts of the same two projects, with the result that both projects suffer delay in the end.

In addition, it is difficult to translate priorities established by management into appropriate action at, for example, the level where material is requisitioned. Yet, the need for priorities to penetrate to the lowest level of the organization is essential if management decisions are to be carried out effectively.

Factors Influencing Project Priorities

While the relative importance of these factors will vary, depending on the organization and the type of project involved, the following list includes most of the factors important in setting priorities (not necessarily in order of importance):

- Completion or delivery date, and its proximity.
- Penalty risks.
- Customer.
- Competitive risks.
- Technical risks.
- Management sponsor.
- Return on investment.
- Magnitude of costs, investment and/or profit.
- Impact on other projects.
- Impact on other affiliated organizations.
- Impact on a particular product line.
- Political and visibility risks.

Project Priority Review Board

For product development, capital investment, and major systems projects, existing procedures should require establishing the relative importance of such projects during the authorization process, since not all projects can usually be funded or supported simultaneously. The priority questions discussed in this section relate to the conflicts that occur after the projects have been authorized or after the award of a contract has been made by a customer.

A Project Priority Review Board has been used in some organizations to deal with this difficult problem. The basic approach of such a board, which is composed of top management people, is as follows:

- Priority factors (such as those listed above) are listed for each project, with a value of 1 to 10 assigned to each factor by several key managers. These scores are summarized and compared.
- Using the comparative scores and other information not easily quantified, the Board meets each month and agrees on the relative priority to be accorded to each project.
- The current priority list, although closely controlled because of its sensitive nature, is distributed to each key manager involved in the projects.

In some organizations, priorities are established by category rather than an absolute ranking of projects. Typically, the "A" category gets top priority, "B" routine priority, and "C" projects are generally on the "backburner" awaiting higher status or the axe.

Lower Level Priority Rules

Experience with many projects in many industries indicates that only about 15 percent of the activities in any one project are truly critical to meeting its completion date. Thus, for a multiproject point of view, the overall project priority discussed in the preceding sections should only be applied when conflicts occur among the truly critical activities of two or more projects.

In order for this approach to be used, it is necessary to plan and schedule each project in such a way that the critical 15 percent of the activities can be identified. Network planning discipline discussed in Chapter 10 is the most effective method of doing this in a consistent manner. When two or more projects are planned in this way, information is provided that allows application of various lower level priority rules in the following way. Other similar priority rules may be developed and used.

Conflict between Activities	Priority Given to:
Critical	Project with highest overall current priority.
Critical versus noncritical	Critical activity (regardless of overall project priority).
Noncritical versus noncritical	Activity with least slack (allowable delay); if equal slack, shortest or longest activity; if same duration, use current project priority. Alternatively, give priority to activity using largest number of critical resources.

Benefits of Using PERT/CPM/PDM[1] Network-Based Project Management Systems in Multiproject Situations

Significant benefits are realized from the proper use of network-based systems in both the multiple major and multiple small project situations.

- Improved planning and scheduling of activities and forecasting of resource requirements.
- Identification of repetitive planning patterns that can be followed in a number of projects, thereby simplifying the planning process.
- Ability to reschedule activities to reflect interproject dependencies and resource limitations, following known priority rules.
- Ability to use the computer effectively to produce timely, valid information for multiproject management purposes.

Interdependencies between and within Projects

Projects and activities within projects can be interrelated in three basic ways.

Result-of-action.   The results produced by completion of an activity in one project or task must be available before an activity in another project or task can begin.

Common-unit-of-resource.   An engineer, for example, must complete an activity in one project or task before he or she can begin another activity in another project or task.

Rate-of-use-of-common-resources.   Two or more projects or tasks are using one resource pool, such as a group of pipe-fitters; when the rate of use of the resource by the projects exceeds the supply, the projects or tasks become interdependent on each other through the limited resource pool.

The first two interdependencies can be represented by interface events as discussed in Part II, and especially in Chapter 13. The third interdependency is treated in more detail in the following section and in Part II.

8.4   RESOURCE MANAGEMENT

A common cause of project delays, resulting in penalties and other undesirable effects, is the overcommitment of the organization to contracts or projects with respect to available resources. Cost overruns can also occur when the limitations of available resources are not considered at the time of project commitment and during project execution. Resource management therefore becomes important to project managers, to those responsible for multiproject situations, and to functional managers contributing to projects. It involves

- Estimating and forecasting the resource requirements by functional task for each project and summarized for all projects.
- Acquiring, providing, and allocating the needed resources in a timely and efficient manner.
- Planning the work for accomplishment within the constraints of limited resources.
- Controlling the use of resources to accomplish the work according to the project plan.

An important difference has been observed in human resource management on one or a few large projects compared with multiple smaller projects. On a large project, the required resources usually will be obtained somehow, either through hiring additional people on a permanent or temporary basis, or by using contractors, consultants or "body shops." On multiple smaller projects, the available, always limited resources usually must be carefully allocated to support all the projects, and it is often more difficult to augment the resource pools with outsiders.

Resources to be Managed

A variety of resources must be managed in this manner, including time, money, people, facilities, equipment, and material. *Time* is a fundamental resource which cannot be managed like the others. Time flows at a constant rate, and time that is not used can never be recovered. It cannot be stored or accumulated for later use. Time is the element that interrelates all other resources with the project plan.

The other resources are forecasted, provided, and controlled in accordance with established procedures in all organizations for the management of departments, divisions, and other organizational elements. However, procedures for performing these resource management tasks on a project basis are frequently inadequate.

Procedures for Resource Management on Projects

Effective procedures for estimating, forecasting, allocating, and controlling the resources on one or many projects are described in Part II. On major projects and in multiproject situations, computer-supported systems are frequently the only practical way to handle the dynamic need to replan, reschedule, and reallocate resources, considering the large volume of detailed information involved.

8.5   MULTIPROJECT OPERATIONS PLANNING AND CONTROL

Situations involving multiple projects, large or small, frequently require the establishment of an operations planning and control function. This may be set up within a division, a product line, or for an entire company. Operations planning and control integrates and controls, on a master plan level, and for all contracts or projects, the functions of marketing, engineering, procurement, manufacturing, and installation, usually within a specific product line or division. Such a function within a particular organization will benefit significantly the organization's project management capabilities, and the organization of the project office, resulting in

- Improved project planning and control support to each project manager, with reduction in planning and control staff in each project office.
- Improved ability to resolve conflicts between projects and to control relative priorities of projects.
- Improved uniformity of project planning and control practices, enabling higher management to review all projects on a more consistent basis.
- Better forecasting of resource requirements for all projects.

Nature of the Problem

The need exists for top management to have confidence that planning is directed towards optimum corporate *and* project performance. It must have adequate feedback on actual compared to forecasted achievement for functional, corporate, and customer performance goals to be able to evaluate the consequences of *specific* cost trade-offs and *specific* alternative business strategies, quickly and quantitatively.

Central planning functions tend to stop short of filling those management needs, because they usually lack the capability to monitor functional planning continuously at a manageable level of detail; evaluate the corporate impact of forecasted functional achievement accurately and routinely; carry out quantitative analyses, within a reasonable time, of the consequences of alternative business strategies; replan, on time, matching available resources with contract requirements; and issue top level schedules reflecting a balanced, coordinated load between engineering, manufacturing, and installation or field operations.

What is needed is a planning capability in a position to evaluate functional plans and their impact on corporate and contract performance objectives, with tools powerful enough to evaluate the effect of discrete contract-oriented decisions on project performance, and so placed in the organizational structure to be able to offer solutions to problems transcending functional organizational boundaries.

Solution

To establish an operations planning and control function with supporting systems to fill these needs. The function should have the following charter:

- Optimization of corporate and project performance goals by coordination of planning activities in the marketing, engineering, manufacturing, and installation functions, through generation of corporate master schedules.
- Coordination of functional planning by means of continuous workload versus capacity evaluation, recognizing functional dependencies in project execution, with a planning horizon extending as far as firm, proposed, and forecasted project activities will allow.
- Continuous evaluation of functional capabilities relative to requirement to (1) allow forecasting of performance against corporate and project goals and (2) highlight areas of deviation where management action is required to resolve cases of potential capacity shortage or surplus.
- Development and maintenance of supporting systems for simulation purposes: to evaluate the likely consequences of alternative business strategies; to offer possible solutions to senior management.

Potential Benefits

The potential benefits of the operations planning and control function are:

- Improving cross-functional communications with a positive effect on functional performance.
- Improving overall contract and project performance.
- Improving corporate performance by improved resource planning and utilization.

Examples of specific benefits include:

- Reduction of project cycle time.
- Increased direct labor efficiency.

- Increased utilization of tools, test equipment, and other facilities.
- Reduction of gross inventory.
- Reduced exposure to contract penalties.

The final effect of all these benefits is to enable the company to handle more projects, and produce more sales revenue and net income without increasing total personnel and capital investment.

Operations Planning and Control Overview

Operations planning and control reflects the application of currently accepted management concepts combined with network based systems. It is the result of viewing the problem outlined above from a management perspective, which recognizes that the proper organizational structure is requisite to the implementation of a planning and control system, and inversely that an organization with inadequate systems and procedures will be ineffective. To solve the problem addressed by operations planning and control, both organizational considerations and systems development must be welded together.

Organizational Considerations

The operations planning and control function must report at a level that ensures complete objectivity and impartiality. The exact placement of the function in a given organization will depend on the company product line structure, and its size. It is clear that the function must report at a director level with responsibility for overall corporate planning, or to the division or product line manager responsible for engineering, manufacturing, and installation.

In those companies with more than one product line, where normal product cycle time exceeds one year, and contributions from two or more functions are required to execute the project, then a separate operations planning and control organization should exist within each product line, and overall coordination should be supplied by a similar function reporting to a corporate level manager.

Operations planning and control performs a distinctly different corporate role from that usually associated with the management of large projects. The project manager is responsible for large one-of-a-kind projects. If he uses a network based system for controlling the project it will be structured to reflect a great amount of detail and will generally *not* be related to other large projects. Operations planning and control on the other hand deals with contracts that reflect a high degree of similarity and are generally smaller than those found under the control of a project manager. The operations planning and control manager can, however, make a valuable contribution to project management by reserving capacity assigned to the project(s), when preparing the master schedule; by providing an effective interface with the functions; and/or by providing most or all of the project planning and control information required for the major projects.

Supporting System

If the operations planning and control organization is to perform its function, systems support will be required for the following:

- Operations scheduling and evaluation.
- Planning evaluation.
- Resource allocation.

Organization and System Interface

The concept of operations planning and control and the interface between the involved organizational functions and the supporting systems are illustrated in Figure 8.2.

The operations planning and control function, reporting to the operating division manager (or equivalent), plans, schedules, monitors, reports, and controls, *at the master schedule level,* all orders and contracts (projects) through all contributing functions.

The set of bridging networks integrates the functional plans and schedules for all projects. The networks are also linked to the supporting functional systems by the downward flow of master scheduling information, and the upward flow of progress and status information. This means that the bridging networks must incorporate milestone or interface events common to both the operations planning and control systems and the functional systems. These milestone/interface events must meet the needs of both types of systems.

Experience indicates that the two elements of operations planning and control—the organizational function and the system—must be developed and introduced concurrently. One element without the other simply will not work. Additionally, if one element is implemented prematurely less than desirable results will be produced.

Multiproject Master Business Planning Using Standard and Unique Milestones

Boznak[2] describes a system for multiproject planning and control based on a four-phased project process (design, procure, build/test, and deliver/support) with a set of standardized milestone events that are scheduled and

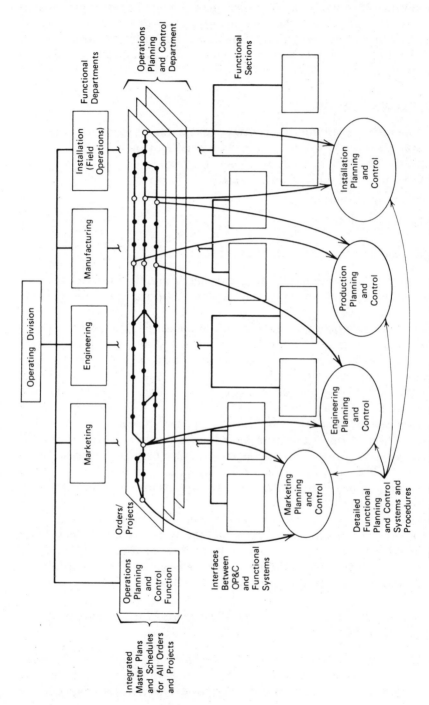

Figure 8.2 General illustration of operations planning and control concept.

controlled by the project manager, and a set of project-unique milestones scheduled and controlled by the functional project leader or manager. The project manager controlled milestones in Boznak's approach are:

- Standardized.
- Coordinative.
- Business-level.
- Summary in nature.

The functional project leader controlled milestones are:

- Project unique.
- Coordinative.
- Functional-level.
- Detailed in nature.

This multiproject system produces several types and levels of reports:

- Customer satisfaction report.
- Multiyear master plan, containing all projects.
- Detail reports of various kinds.
- Management action reports, listing all milestones not on schedule.
- Monthly performance summary, showing schedule compliance by project and functional organization.

There are a number of ways to plan, schedule, track, and control multiple projects within an organization. The key to success is to have a coherent, integrated, consistent system for doing this, and to make sure that everyone is using the same system.

PART
II

‹ ›

Managing Specific Projects

Chapters 9 through 16 contain specific, detailed information for guidance and use in establishing practices and procedures for specific projects. Since it is not practical to provide examples related to every type of project of interest to all readers, the emphasis in Part II is on major commercial and product development projects in manufacturing companies. However, the material can readily be adapted to other types of projects in other environments.

For simplicity, the word "project" will be used throughout Part II, although in most instances "program" can be used interchangeably. These terms are defined in Chapter 2.

9

❮ ❯

Organizing the Project Office and Project Team

The approach described in this and the following chapters is based on typical major project situations involving the design, manufacture, assembly, and testing of complex hardware and software systems. This presumes the following conditions:

- The project (or program) warrants a full-time project manager.
- The project office is held to minimum size, with maximum use of functional contributors in existing departments.

This chapter summarizes the functions of the project office and the project team under these conditions, describes the duties of key persons involved in the project, and discusses their relationships.

As discussed in Chapter 8, the situation frequently occurs wherein one project manager is responsible for several projects, or, when multiple small projects exist, the division or other general manager retains the project manager responsibility himself. In still other situations, a manager of projects is appointed. In such cases, a centralized Operations Planning and Control function is usually required, as described in section 8.6.

9.1 FUNCTIONS OF THE PROJECT OFFICE AND PROJECT TEAM

The *project office* supports the project manager in carrying out his or her responsibilities. Thus the project manager's basic charter, organizational relationships, and the nature of the project itself will influence the makeup of the project office. The presence or absence of other projects, and of a central planning office, will also affect the organization of the project office.

The *project team* includes all functional contributors to the project, as well as the members of the project office. The general functions to be carried out during completion of the overall project by members of the project team are:

- Management.
- Product design and development.
- Product manufacture.
- Product installation, testing and field support.

The relationships of these functions to the project manager are shown in the generalized project organization chart in Figure 3.2. Each of these functions is discussed in the following paragraphs.

Management

The management functions are those functions necessary to enable the project manager to fulfill basic responsibilities: Overall direction and co-ordination of the project through all its phases to achieve the desired results within established budget and schedule. These functions are summarized in Chapter 3 and discussed in detail in the remainder of the book.

Product Design and Development

The basic purpose of this general function is to produce adequate documentation (and often a prototype product or system) so that the product may be manufactured in the quantity required within the desired cost and schedule. These functions may be defined as

- Systems analysis, engineering, and integration
- Product design
- Product control (quality, cost, configuration)

Systems analysis, engineering, and integration functions include: system studies; functional analysis and functional design of the system or product; and coordination and integration of detailed designs, including functional and mechanical interfaces between major subsystems or components of the product.

Product design functions include the detailed engineering design and development functions needed to translate the functional systems design into specifications, drawings, and other documents which can be used to manufacture, assemble, install, and test the product. This may also include the manufacture and test of a prototype or first article system or product, using either model shop or factory facilities on a subcontract basis.

Product control functions include: product quality control, using established staff specialists and procedures; product cost control, including value engineering practices; product configuration control including design freeze practices (to establish the "base line" design), engineering change control practices, and documentation control practices.

The term "product" refers to *all* results of the project: hardware, software, documentation, training or other services, facilities, and so on. A new organizational entity can also be a result of the project.

The project office in a specific situation may perform none, a few, or all of these product design and development functions, depending on many factors. Generally, a larger share of these functions will be assigned to the project office (together with adequate staff) when the product is new or unusual to the responsible unit, or when there is little confidence that the work will be carried out efficiently and on schedule within established engineering departments. When several engineering departments, for example, from different product lines or different companies, are contributing to the product design and development, the functions of systems analysis, engineering and integration, and of product control, should be assigned to the project office. Figure 3.3 illustrates a project with a large technical staff assigned to the project office for an electronic telephone switching system contract.

Except in the situation described above, these functions are usually performed by project team members within existing engineering departments, under the active coordination of the project manager.

Product Manufacture

This general function is to purchase materials and components, fabricate, assemble, test, and deliver the equipment required to complete the project. These functions are carried out by the established manufacturing departments within the project's parent company or by outside companies on a subcontract or purchase order basis.

The project manager, however, must coordinate and integrate the manufacturing functions with product design and development on one hand and field operations (if any) on the other. *The lack of proper integration between functions is the most common cause of project failure.*

In order to achieve this integration, it is necessary to appoint a project manufacturing coordinator or equivalent who will, in effect, act as a project manager for product manufacture. This person is a key project team member, and may be assigned full time to one major project or may be able to handle two or more projects at one time, if the projects are small.

It is recommended that the project manufacturing coordinator remain within the appropriate manufacturing department. When two or more divisions or companies perform a large part of the product manufacture, each must designate a project manufacturing coordinator, with one designated as the lead division for manufacture. If it is not possible to designate a lead division, then the coordination effort must be accomplished by the project office.

Purchasing and Subcontracting

This function is sometimes included with the product manufacture area, but it is normally important enough to warrant full functional responsibility.

A separate project purchasing and subcontracting coordinator with status equivalent to that of the manufacturing coordinator should be appointed to handle all purchasing and subcontracting matters for the project manager. This person should be a part of the purchasing department to maintain day-to-day contact with all persons carrying out the procurement functions.

Product Installation, Testing, and Field Support

Many projects require field installation and testing, and some include continuing field support for a period of time. In these cases, a field project manager (or equivalent) is required.

When field operations are a part of the project, this phase is usually clearly recognized as being of a project nature and requiring one person to be in charge. This field project manager is almost always a member of an established installation department (or equivalent) if such a department exists within the responsible company. Since engineering and manufacturing operations frequently overlap the installation phase, the overall project manager's role continues to be of critical importance to success while field operations are in progress. However, in the relationship between the project manager and the field project manager, the project manager retains the overall responsibility for the coordination of the entire project.

Assignment of Persons to the Project Office

As a general rule, it is recommended that the number of persons assigned to a project office under the supervision of the project manager be kept as small as possible. This will emphasize the responsibility of each functional (line) department or staff for their contribution to the project and retain to the maximum degree the benefits of specialized functional departments. It will also increase flexibility of functional staffing of the project, avoid unnecessary payroll costs to the project, and minimize reassignment problems when particular tasks are completed. This will enable the project manager to devote maximum effort to the project itself, rather than supervisory duties related to a large staff.

With adequate project planning and control procedures, a qualified project staff can maintain the desired control of the project. In the absence of adequate planning and control procedures to integrate the functional contributions, it is usually necessary to build up a larger staff with as many functional contributors as possible directly under the project manager in order to achieve control. This is an expensive and frequently awkward approach, and it aggravates the relationships between the project manager and contributing functional managers.

The persons who should be assigned (transferred) permanently to the project office are those who:

- Deal with the management aspects of the project.
- Are needed on a full-time basis for a period of at least 6 months.
- Must be in frequent close contact with the project manager or other members of the project office in the performance of their duties.
- Cannot otherwise be controlled effectively, because of organizational or geographic considerations.

The recommended assignment location of each of the key people on the project team follows:

Project manager.   The project manager is always considered the manager of the project office (which could be a one-person office).

Project engineer.   The project engineer may be assigned to the project office in charge of product design and development where the product is new to the company or where several divisions are involved, as discussed earlier. Otherwise, the project engineer should remain within the lead engineering department.

Contract administrator.   The contract administrator should remain a member of the contract administration staff, except on very large programs extending over a considerable period of time, but may be located physically in the project office while remaining with his or her parent organization.

Project controller.   The project controller should always be assigned to the project office, except where not needed full time or where a centralized planning and control function adequately serves the project manager.

Project accountant.   The project accountant should remain a member of the accounting department, except on very large programs extending over a considerable period of time. Like the contract administrator, she may physically be located in the project office while remaining with her parent organization.

Manufacturing coordinator.   The manufacturing coordinator should remain a member of the manufacturing organization, preferably on the staff of the manufacturing manager or within production control. When more than one division is to contribute substantially to product manufacture, it may be necessary to assign this person to the project office to enable effective coordination of all contributors.

Purchasing and subcontracting coordinator.   This coordinator should remain a member of the purchasing department in most cases.

Field project manager.   The field project manager should remain a member of the installation or field operations department, if one exists, except under unusual circumstances that would require him to be assigned to the project office.

Project Team Concept

As discussed in Chapter 5, whether a person is assigned to the project office or remains in a functional department or staff, all persons holding identifiable responsibilities for direct contributions to the project are considered to be members of the project team. Creating awareness of membership in the project team is a primary task of the project manager, and development of a good project team spirit has proven to be a powerful means for accomplishing difficult objectives under tight time schedules. Outside contributors (vendors, contractors, consultants, architects, etc.) are also members of the project team, as are contributing customer representatives.

The Project Organization Chart

Figure 3.14 shows a typical representation of a project team in the format of a classic organizational chart. This type of representation can be confusing if not properly understood, but it can also be useful to identify the key project team members and show their relationships to each other and to the project manager *for project purposes.* This recognizes that such a chart does not imply permanent superior-subordinate relationships portrayed in the company organization charts.

9.2   PROJECT MANAGER DUTIES

Within the general statement of the project manager responsibilities given in Chapter 4, the following description of project manager duties is presented as a guide for development of specific duties on a particular project. Some of the duties listed may not be practical, feasible, or pertinent in certain cases, but wherever possible it is recommended that all items mentioned be included in the project manager's duties and responsibilities, with appropriate internal documentation and dissemination to all concerned managers.

Project Start-Up

- Identify key project team members and define their responsibilities.
- Rapidly and efficiently plan and start up the project (see Chapter 11 and Appendix B).

General

- Assure that all equipment, documents, and services are properly delivered to the customer for acceptance and use within the contractual schedule and costs.
- Convey to all concerned departments (both internal and external) a full understanding of the customer requirements of the project.
- Participate with and lead responsible managers and key team members in developing overall project objectives, strategies, budgets, and schedules.
- Plan for all necessary project tasks to satisfy customer and management requirements, and assure that they are properly and realistically scheduled, budgeted, provided for, monitored, and reported.
- Identify promptly all deficiencies and deviations from plan.
- Assure that actions are initiated to correct deficiencies and deviations, and monitor execution of such actions.
- Assure that payment is received in accordance with contractual terms.

- Maintain cognizance over all project contacts with the customer and assure that proper team members participate in such contacts.
- Arbitrate and resolve conflicts and differences between functional departments on specific project tasks or activities.
- Maintain day-to-day liaison with all functional contributors to provide communication required to assure realization of their commitments.
- Make or force required decisions at successively higher organizational levels to achieve project objectives, following agreed escalation procedures.
- Maintain communications with higher management regarding problem areas and project status.

Customer Relations

In close cooperation with the customer relations or marketing department:

- Receive from the customer all necessary technical, cost, and scheduling information required for accomplishment of the project.
- Establish good working relationships with the customer on all levels: management, contracts, legal, accounts payable, system engineering, design engineering, field sites, and operations.
- Arrange and attend all meetings with customer (contractual, planning, engineering, operations).
- Receive and answer all technical and operational questions from the customer, with appropriate assistance from functional departments.

Contract Administration

- Identify any potential areas of exposure in existing or potential contracts and initiate appropriate action to alert higher management and eliminate such exposure.
- Prepare and send, or approve prior to sending by others, all correspondence on contractual matters.
- Coordinate the activities of the project contract administrator in regard to project matters.
- Prepare for and participate in contract negotiations.
- Identify all open contractual commitments.
- Advise engineering, manufacturing, and field operations of contractual commitments and variations allowed.
- Prepare historical or position papers on any contractual or technical aspect of the program, for use in contract negotiations or litigation.

Project Planning, Control, Reporting, Evaluation, and Direction

- Perform, or supervise the performance of, all project planning, controlling, reporting, evaluation and direction functions as described in following chapters, as appropriate to the scope of each project.
- Conduct frequent, regular project evaluation and review meetings with key project team members to identify current and future problems and initiate actions for their resolution.
- Prepare and submit weekly or monthly progress reports to higher management, and to the customer if required.
- Supervise the project controller and his staff.

Marketing

Maintain close liaison with marketing and utilize customer contacts to acquire all possible marketing intelligence for future business.

Engineering

- Insure that engineering fulfills its responsibilities for delivering, on schedule and within product cost estimates, drawings and specifications usable by manufacturing and field operations, meeting the customer specifications.
- In cooperation with the engineering, drafting, and publications departments, define and establish schedules and budgets for all engineering and related tasks. After agreement, release funding allowables and monitor progress on each task in relation to the overall project.
- Act as the interface with the customer for these departments (with their assistance as required).
- Assure the control of product quality, configuration, and cost.
- Approve technical publications prior to release to the customer.
- Coordinate engineering support related to the project for manufacturing, installation, legal, and other departments.
- Participate (or delegate participation) as a voting member in the Engineering Change Control Board on matters affecting the project.

Manufacturing

- Insure that manufacturing fulfills its responsibilities for on-schedule delivery of all required equipment, meeting the engineering specifications within estimated manufacturing costs.

- Define contractual commitments to production control.
- Develop schedules to meet contractual commitments in the most economical fashion.
- Establish and release manufacturing and other resource and funding allowables.
- Approve and monitor production control schedules.
- Establish project priorities.
- Approve, prior to implementation, any product changes initiated by manufacturing.
- Approve packing and shipping instructions based on type of transportation to be used and schedule for delivery.

Purchasing and Subcontracting

- Insure that purchasing and subcontracting fulfills their responsibilities to obtain delivery of materials, equipment, documents, and services on schedule and within estimated cost for the project.
- Approve make-or-buy decisions for the project.
- Define contractual commitments to purchasing and subcontracting.
- Establish and release procurement funding allowables.
- Approve and monitor major purchase orders and subcontracts.
- Specify planning, scheduling, and reporting requirements for major purchase orders and subcontracts.

Installation, Construction, Testing, and Other Field Operations

- Insure that installation and field operations fulfill their responsibilities for on-schedule delivery to the customer of materials, equipment, and documents within the cost estimates for the project.
- Define contractual commitments to installation and field operations.
- In cooperation with installation and field operations, define and establish schedules and budgets for all field work. After agreement, release funding allowables and monitor progress on each task in relation to the overall project.
- Coordinate all problems of performance and schedule with engineering, manufacturing, and purchasing and subcontracting.
- Except for customer contacts related to daily operating matters, act as the customer interface for installation and field operations departments.

Financial

In addition to the financial planning and control functions described in later chapters:

- Assist in the collection of accounts receivable.
- Approve prices of all change orders and proposals to the customer.

Project Closeout (See Appendix B)

- Insure that all required steps are taken to present adequately all project deliverable items to the customer for acceptance and that project activities are closed out in an efficient and economical manner.
- Assure that the acceptance plan and schedule comply with the customer contractual requirements.
- Assist the legal, contract administration, and marketing or commercial departments in preparation of a closeout plan and required closeout data.
- Obtain and approve closeout plans from each involved functional department.
- Monitor closeout activities, including disposition of surplus materials.
- Notify finance and functional departments of the completion of activities and of the project.
- Monitor payment from the customer until all collections have been made.

9.3 FUNCTIONAL PROJECT LEADER DUTIES

The functional project leaders direct, lead, and integrate the activities of the project team members (all persons contributing to the project) within their specific function. This manager may be a senior functional manager, but it is usually recommended that this role be delegated to a subordinate who can devote the required amount of time to the project leader job. If the assignment is for a large project requiring a full-time functional project leader, it obviously cannot be a person who also has responsibility for managing the functional department. Frequently, the assignment is given to a person who is also directly responsible for carrying out one or more specific tasks within the functional department in question, causing potential conflicts due to the project leader's favoring the tasks for which they are responsible. This is a dilemma that is not easy to solve; some organizations have given the project leader role for more than one project (none of which requires a full-time assignment) to the same person.

The specific duties of the functional project leader will vary considerably, depending on the nature of the particular function that person represents. The project leader is really a mini-project manager, and many of the project manager's duties described above can be translated to the functional project leader, with appropriate limitations on the scope of activity. A well-established functional project leader role is that of the project engineer, whose duties are described in some detail in the following section. These duties, and those of the other team members following, can also be used as a guide to defining other functional project leaders' duties, again with appropriate translation to fit the specific function.

9.4   PROJECT ENGINEER DUTIES

General

The project engineer is responsible for the technical integrity of the project and for cost and schedule performance of all engineering phases of the project. Specifically, the responsibilities of the project engineer are:

- Insure that the customer performance requirements are fully understood and that the company is technically capable of meeting these requirements.
- Define these requirements to the smallest subsystem to the functional areas so that they can properly schedule, cost, and perform the work to be accomplished.
- Insure that the engineering tasks so defined are accomplished within the engineering schedules and allowables (manpower, materials, funds) of the contract.
- Provide technical direction as necessary to accomplish the project objectives.
- Conduct design review meetings at regular intervals to assure that all technical objectives will be achieved.
- Act as technical advisor to the project manager and other functional departments, as requested by the project manager.

In exercising the foregoing responsibilities, the project engineer is supported by the various engineering departments through the designated engineering project leaders.

Proposal Preparation and Negotiation

During the proposal phase, the project engineer will do the following:

- Coordinate and plan the preparation of the technical proposal.
- Review and evaluate the statement of work and other technical data.
- Establish an engineering proposal team or teams.
- Within the bounds of the overall proposal schedule, establish the engineering proposal schedule.
- Reduce customer engineering requirements to tasks and subtasks.
- Define in writing the requirements necessary from engineering to other functional areas, including preliminary specifications for make or buy, or subcontract items.
- Coordinate and/or prepare a schedule for all engineering functions, including handoff to and receipt from manufacturing.
- Review and approve all engineering subtask and task costs, schedules, and narrative inputs.
- Coordinate and/or prepare overall engineering cost.
- Participate in preliminary make-or-buy decisions.
- Participate in overall cost and schedule review.
- Participate, as required, in negotiation of contract.
- Bring problems between the project engineer and engineering functional managers and project leaders to appropriate engineering directors for resolution.

Project Planning and Initiation

The project engineer is responsible for the preparation of plans and schedules for all engineering tasks within the overall project plan established by the project manager. In planning the engineering tasks, he or she will compare the engineering proposal against the received contract. Where the received contract requirements dictate a change in cost, schedule, or technical complexity for solution, he will obtain approval from the director of engineering and the project manager to make the necessary modifications in engineering estimates of the proposal. During this phase, the project engineer will:

- Update the proposal task and subtask descriptions to conform with the contract, and within the engineering allowables prepare additional tasks and subtasks as required to provide a complete engineering implementation plan for the project.

- Prepare a master engineering schedule in accordance with the contractual requirements.
- Prepare, or have prepared, detailed task and subtask definitions and specifications. Agree on allowables, major milestones, and evaluation points in tasks with the task leaders and their functional managers.
- Through the functional engineering managers or project leaders, assign responsibility for task and subtask performance, and authorize the initiation of work against identified commitments based on cost and milestone schedules, with approval of the project manager.
- Using contract specifications as the base line, prepare, or have prepared, specifications for subcontract items.
- Participate and provide support from appropriate engineering functions in final make-or-buy decisions and source selection.
- Prepare, or have prepared, hardware and system integration and acceptance test plan. Review the test plan with Quality Assurance and advise them as to the required participation of other departments.

Project Performance and Control

The project engineer is responsible for the engineering progress of the project and compliance with contract requirements, cost allowables, and schedule commitments. Within these limits, the project engineer, if necessary, may make design changes and task requirement changes in accordance with the project concept and assume the responsibility for the change in concert with the functional engineering project leaders and with the knowledge of the project manager. No changes may be made that affect other functional departments without the knowledge of that department, documentation to the project manager, and the inclusion of the appropriate charge-back of any variance caused by change. The project engineer maintains day-to-day liaison with the project manager for two-way information exchange. Specific responsibilities of the project engineer are to

- Prepare and maintain a file of all project specifications related to the technical integrity and performance.
- Prepare and maintain updated records of the engineering expenditures and milestones and conduct regular reviews to insure engineering performance as required.
- Initiate and prepare new engineering costs-to-complete reports as required.
- Establish work priorities within the engineering function where conflict exists: arbitrate differences and interface problems within the engineering function, and request through functional managers changes in personnel assignments if deemed necessary.

- Plan and conduct design review meetings and design audits as required, and participate in technical reviews with customer.
- Prepare project status reports as required related to engineering.
- With the project manager and other functional departments, participate in evaluation and formulation of alternative plans as required by schedule delays or customer change requests.
- Assure support to purchasing and subcontracting, manufacturing, field operations, and support activities by providing liaison and technical assistance within allowables authorized by the project manager.
- Modify and reallocate tasks and subtasks, open and close cost accounts, and change allowable allocations within the limits of the approved engineering allowables, with the concurrence of the functional managers involved. Provide details to the project manager of all such actions prior to change.
- As requested by the project manager, support legal and contracts administration by providing technical information.
- Review and approve technical aspects of reports for dissemination to the customer.
- Authorize within the approved allowables the procurement of material and/or services as required for the implementation of the engineering functional responsibility.
- Adjudicate technical problems and make technical decisions within scope of contractual requirements. Cost and schedule decisions affecting contractual requirements or interfaces with other functions are to be approved by the appropriate engineering function manager with the cognizance of the director of engineering (or his delegate) and the project manager.
- Approve all engineering designs released for procurement and/or fabrication for customer deliverable items.
- Bring problems arising between the project engineer and engineering functional managers to the engineering director for resolution.
- Bring problems arising between the project engineer and functions outside engineering to the project manager for resolution, with the cognizance of the director of engineering and the director of the other functions.

9.5 CONTRACT ADMINISTRATOR DUTIES

General

Contract administration is a specialized management function indispensable to effective management of those projects carried out under contract with customers. This function has many legal implications and serves to protect the company from unforeseen risks prior to contract approval and during execution of the project. Well-qualified, properly organized contract administration support to a project manager is vital to the continuing success of companies responsible for major sales contracts.

Contract administration is represented both on the project manager's team and on the general manager's staff. A director of contract administration has the authority to audit project contract files and to impose status reporting requirements that will disclose operational and contractual problems relating to specific projects. The director of contract administration is also available to provide expertise in the resolution of contract problems beyond the capability of the contract administrator assigned to a given project.

The project contract administrator is responsible for day-to-day administration of (1) the contract(s) that authorize performance of the project and (2) all subcontracts with outside firms for equipment, material, and services to fulfill project requirements.

Proposal Preparation

- When participation of an outside subcontractor is required, assure that firm quotations are obtained based on terms and conditions compatible with those imposed by the customer.
- Review with the legal and financial departments all of the legal and commercial terms and conditions imposed by the customer.
- Review the proposal prior to submittal to assure that all risks and potential exposures are fully recognized.

Contract Negotiation

- Lead all contract negotiations for the project manager.
- Record detailed minutes of the proceedings.
- Assure that all discussions or agreements reached during negotiations are confirmed in writing with the understanding that they will be incorporated into the contract during the contract definition phase.
- Assure that the negotiating limits established by the Proposal Review Board (or equivalent) are not exceeded.

Contract Definition

- Expedite the preparation, management review, and execution of the contract, as follows:

Clarify the contract format with the customer.

Establish the order of precedence of contract documents incorporated by reference.

Set the date by which the contract will be available in final form for management review prior to execution.

Participate with the project manager in final briefing of management on the contract terms and conditions prior to signature.

Project Planning Phase

- Establish channels of communication with the customer and define commitment authority of project manager, contract administrator, and others.
- Integrate contract requirements and milestones into the project plan and schedule, including both company and customer obligations.
- Establish procedures for submission of contract deliverables to customer.
- Establish mechanics for monthly contract status reports for the customer and management.

Project Execution Phase

- Monitor and follow up all contract and project activities to assure fulfillment of contractual obligations by both the company and the customer.
- Assure that all contract deliverables are transmitted to the customer and that all contractually required notifications are made.
- Record any instance where the customer has failed to fulfill his obligations and define the cost and schedule impact on the project of such failure.
- Identify and define changes in scope, customer-caused delays, and force majeure, including:

  Early identification and notification to customer.

  Obtaining customer's agreement that change of scope or customer-caused delay or force majeure case has actually occurred.

  In coordination with the project manager and the project team, prepare a proposal that defines the scope of the change(s) and resulting price and/or schedule impact for submittal to the customer for eventual contract modification.
- Assist in negotiation and definition of contract change orders.

- Participate in project and contract status reviews and prepare required reports.
- Arrange with the customer to review the minutes of joint project review meetings to assure that they accurately reflect the proceedings.
- Assure that the customer is notified in writing of the completion of each contractual milestone and submission of each contract deliverable item, with a positive assertion that the obligation has been fulfilled.
- Where the customer insists on additional data or work before accepting completion of an item, monitor compliance with his requirement to clear such items as quickly as possible.

Project Closeout Phase (See Appendix B)

- At the point where all contractual obligations have been fulfilled, or where all but longer term warranties or spare parts deliveries are complete, assure that this fact is clearly and quickly communicated in writing to the customer.
- Assure that all formal documentation related to customer acceptance as required by the contract is properly executed.
- Expedite completion of all actions by the company and the customer needed to complete the contract and claim final payment.
- Initiate formal request for final payment.
- Where possible, obtain certification from the customer acknowledging completion of all contractual obligations and releasing the company from further obligations, except those under the terms of guaranty or warranty, if any.

Project/Contract Record Retention

Prior to disbanding the project team, the project contract administrator is responsible for collecting and placing in suitable storage the following records, to satisfy legal and internal management requirements:

- The contract file, which consists of:
    Original request for proposal (RFP) and all modifications.
    All correspondence clarifying points in the RFP.
    Copy of company's proposal and all amendments thereto.
    Records of negotiations.
    Original signed copy of contract and all documents and specifications incorporated in the contract by reference.

All contract modifications (supplemental agreements).

A chronological file of all correspondence exchanged between the parties during the life of the program. This includes letters, telexes, records of telephone calls, and minutes of meetings.

Acceptance documentation.

Billings and payment vouchers.

Final releases.

- Financial records required to support postcontract audits, if required by contract or governing statutes.
- History of the project (chronology of all events—contractual and noncontractual).
- Historical cost and time records that can serve as standards for estimating future requirements.

9.6   PROJECT CONTROLLER DUTIES

The primary responsibility of the project manager is to plan and control his project. On some smaller or less complex projects, he may be able to perform all the planning and controlling functions himself. However, on most major projects, as defined in Chapter 2, it will be necessary to provide at least one person who is well qualified in project planning and control and who can devote full attention to these specialized project management needs. This person is the project controller. (A number of other equivalent job titles are in use for this position.) On very large or complex programs or projects the project controller may require one or more persons to assist in carrying out these duties and responsibilities.

If a centralized operations planning and control function exists in the company (discussed in Chapter 8), that office may provide the needed planning and control services to the project manager. In that case, the project controller would be a member of the Operations Planning and Control Office and would have available the specialists in that office. In other situations the project controller may be transferred from Operations Planning and Control to the project office for the duration of the project.

The duties of the project controller are described in the following sections.

General

- Perform for the project manager the project planning, controlling, reporting, and evaluation functions (described in following chapters) as delegated to him, so that the project objectives are achieved within the schedule and cost limits.

- Assist the project manager to achieve clear visibility of all contract tasks so that they can be progressively measured and evaluated in sufficient time for corrective action to be taken.

Project Planning and Scheduling

- In cooperation with responsible managers define the project systematically so that all tasks to be performed are identified and hierarchically related to each other, including work funded under contract or by the company, using the project breakdown structure or similar technique.
- Identify all elements of work (tasks or work packages) to be controlled for time, manpower, or cost, and identify the responsible and performing organizations and project leaders for each.
- Define an adequate number of key milestones for master planning and management reporting purposes.
- Prepare and maintain a graphic project master plan and schedule, based on the project breakdown structure, identifying all tasks or work packages to be controlled in the time dimension, and incorporating all defined milestones.
- Prepare more detailed graphic plans and schedules for each major element of the project.

Budgeting and Work Authorization

- Obtain from the responsible functional project leader for each task or work package a task description, to include:

    Statement of work.

    Estimate of resources required (work days, computer hours, etc.).

    Estimate of labor, computer, and other costs (with assistance of the project accountant).

    Estimate of start date, and estimated total duration and duration between milestones.

- Prepare and maintain a task description file for the entire project.
- Summarize all task manpower and cost estimates, and coordinate needed revisions with responsible managers and the project manager to match the estimates with available and allocated funds for the project in total, for each major element, and for each task.

- Prepare and release, on approval of the project manager and the responsible functional manager, work authorization documents containing the statement of work, budgeted labor, and cost amounts; scheduled dates for start, completion, and intermediate milestones; and the assigned cost accounting number.
- Prepare and release, with approval of the project manager, revised work authorization documents when major changes are required or have occurred, within the authorized funding limits and the approval authority of the project manager.

Work Schedules

- Assist each responsible manager or project leader in developing detailed plans and schedules for assigned tasks, reflecting the established milestone dates in the project master plan.
- Issue current schedules to all concerned showing start and completion dates of all tasks and occurrence dates of milestones.

Progress Monitoring and Evaluation

- Obtain weekly reports from all responsible managers and project leaders of

  Activities started and completed.

  Milestones completed.

  Estimates of time required to complete activities or tasks under way.

  Changes in future plans.

  Actual or anticipated delays, additional costs, or other problems that may affect other tasks, the schedule, or project cost.
- Record reported progress on the project master plan and analyze the effect of progress in all tasks on the overall project schedule.
- Identify major deviations from schedule and determine, with the responsible managers and the project manager, appropriate action to recover delays or take advantage of early completion of tasks.
- Obtain monthly cost reports and compare to the estimates for each current task, with summaries for each level of the project breakdown structure and the total project.
- Through combined evaluation of schedule and cost progress compared to plan and budget, identify deviations that require management action and report these to the project manager.

- Participate in project review meetings, to present the overall project status and evaluate reports from managers and project leaders.
- Record the minutes of project review meetings and follow up for the project manager all resulting action assignments to assure completion of each.
- Advise the project manager of known or potential problems requiring his attention.
- Each month or quarter obtain from each responsible manager an estimate of time, manpower, and cost to complete for each incomplete task or work package; and prepare, in cooperation with the project accountant, a revised projection of cost to complete the entire project.

Schedule and Cost Control

- When schedule or budget revisions are necessary, due to delay or changes in the scope of work, prepare, negotiate, and issue new project master plans, schedules and revised work authorization documents, with approval of the project manager, within the authorized funding limits and the approval authority of the project manager.
- In coordination with the project accountant, notify the Finance Department to close each cost account and reject further charges when work is reported complete on the related task.

Reporting

- Prepare for the project manager monthly progress reports to management and the customer.
- Provide cost-to-complete estimates and other pertinent information to the project accountant for use in preparing contract status reports.
- Prepare special reports as required by the project manager.

9.7   PROJECT ACCOUNTANT DUTIES

The basic function of the project accountant is to provide to the project manager the specialized financial and accounting assistance and information needed to forecast and control manpower and costs for the project. The project accountant duties are as follows:

- Establish the basic procedure for utilizing the company financial reporting and accounting system for project control purposes to assure that all costs are properly recorded and reported.

- Assist the project controller in developing the project breakdown structure to identify the tasks or project elements that will be controlled for manpower and cost.

- Establish account numbers for the project and assign a separate number to each task or work element to be controlled.

- Prepare estimates of cost, based on manpower and other estimates provided by the controller, for all tasks in the project when required to prepare revised estimates to complete the project.

- Obtain, analyze, and interpret labor and cost accounting reports, and provide the project manager, project controller, and other managers in the project with appropriate reports to enable each to exercise needed control.

- Assure that the information being recorded and reported by the various functional and project departments is valid, properly charged, and accurate, and that established policies and procedures are being followed for the project.

- Identify current and future deviation from budget of manpower or funds, or other financial problems, and in coordination with the project controller notify the project manager of such problems.

- Prepare, in coordination with the project manager and the project controller, sales contract performance reports as required by division or company procedures on a monthly basis for internal management purposes, and for submission to any higher headquarters.

9.8   MANUFACTURING COORDINATOR DUTIES

General

The general duties and responsibilities of the manufacturing coordinator (sometimes called the project leader—manufacturing) are to plan, implement, monitor, and coordinate the manufacturing aspects of an assigned project (or projects, where it is feasible to coordinate more than one contract).

Specific Duties

- Review all engineering releases before acceptance by manufacturing to insure they are complete and manufacturable (clean releases), and that all changes are documented by a formal written engineering change request.

- Participate in the development of project master schedules during proposal, negotiation, and execution phases, with particular emphasis on determination of requirements for engineering releases, critical parts lists, equipment requirements, and so on, to insure meeting delivery requirements.

- Monitor all costs related to assigned projects to assure adherence to manufacturing costs and cost schedules. Analyze variances and recommend corrective action. Collect needed information and prepare manufacturing cost to complete.

- Develop or direct the development of detailed schedules for assigned projects, coordinating the participation of manufacturing and product support engineering, material planning, fabrication, purchasing, material stores, assembly, test, quality control, packing and shipping, in order to insure completion of master project schedule within budget limits; provide information and schedules to different functional groups in order for action to be initiated.

- Approve all shipping authorizations for assigned projects.

- Provide liaison between the project manager and Manufacturing; diligently monitor manufacturing portions of assigned projects and answer directly for manufacturing performance against schedules; prepare status reports and provide information needed to prepare costs to complete as required.

- Take action within area of responsibility and make recommendations for corrective action in manufacturing areas to overcome schedule slippages; obtain approval from the project manager for incurring additional manufacturing costs.

- Coordinate requests for clarifications of the impact of contract change proposals on manufacturing effort.

- Participate in the preparation and approval of special operating procedures.

- Review and approve for manufacturing all engineering releases and engineering change notices affecting assigned projects, and participate in Change Control Board activity.

- Represent project manager on all Make/Buy Committee actions.

9.9   FIELD PROJECT MANAGER DUTIES

General

The field project manager (or equivalent) has overall responsibility for constructing required facilities and installing, testing, maintaining for the specified time period, and handing over to the customer all installed equipment and related documentation as specified by the contract. This includes direct supervision of all company and subcontractor field personnel, through their respective managers or supervisors.

Specific Duties

- Participate in the development of project master schedules during proposal, negotiation, and execution phases, with particular emphasis on determination of equipment delivery schedules and manpower and special test equipment needs.

- Monitor all field operations costs for the project to assure adherence to contract allowables. Analyze variances and recommend corrective actions. Collect needed information and prepare field operations cost to complete.

- Develop or direct the development of detailed schedules for all field operations: coordinating the equipment delivery schedules from Manufacturing and subcontractors with field receiving, inspection, installation, testing, and customer acceptance procedures, with due regard for transportation and import/export requirements, to insure completion of the master project schedule within budget limits; provide information and schedules to different functional groups or departments in order for action to be initiated.

- Provide liaison between the the project manager and installation and field operations; diligently monitor field operations portion of the project and answer for performance against schedules; prepare status reports.

- Take action and make recommendations for corrective action in field operations and other areas to overcome schedule slippages; obtain approval of the project manager for incurring additional installation costs.

- Coordinate requests for clarifications of the impact of contract change proposals on field operations.

These model statements of responsibilities can be used to develop additional statements for other project team members.

10

‹ ›

Planning Projects

The triad of project management concepts introduced in Chapter 1 consists of (1) identified points of integrative project responsibility (the project sponsor, the project manager, and the functional project leaders); (2) integrative and predictive project planning and control systems; and (3) the project team. This triad of project management concepts are interdependent. The project manager and the functional project leaders must direct the planning and control efforts, understand and believe in the systems and tools used, and actively use the results. If this occurs, good project teamworking will result. In this and subsequent chapters, we will describe project planning and control: the related functions, tools methods and systems, and discuss how these are used by the project manager and project team members.

10.1 THE PROJECT MANAGER'S PLANNING AND CONTROL RESPONSIBILITIES

Effective project management requires having good planning, scheduling, estimating, budgeting, work authorization, monitoring, reporting, evaluation, and control methods and procedures. It requires not only having such methods and procedures in place, but also requires that the project manager

- Understand and actually use these methods and procedures.
- Perform key planning work at the master schedule level, and give adequate direction to those who perform the detailed planning and control work.
- Establish and maintain effective control of the project.
- Assure that all plans and schedules are adequate and valid.
- Assure that the planning and control functions, as described in the various position descriptions in Chapter 9, are performed properly.

Integrated Planning and Control

Project control is established by

- Mutually setting objectives and goals.
- Defining the tasks to be performed.
- Planning and scheduling the tasks on the basis of required and available resources.
- Measuring progress and performance through an established, orderly system.
- Taking proper corrective action by each project contributor when progress is not made according to plans, or when plans must be changed.
- Resolving schedule and resource conflicts and raising unresolved conflicts to successively higher management levels until a resolution is reached.

Integrated project planning and control means putting together all essential elements of information related to the products or results of the project, time or schedules, and cost, in money, manpower, or other vital resources for all (or as many as practical) life-cycle phases of the project.

The objective of the planning and control effort is to document current plans, schedules, and budgets; compare actual results with each of these; and continually forecast the total project time and cost at completion to enable evaluation, the making of proper decisions, and follow-up of the effect of the decisions.

10.2 PROJECT PLANNING AND THE PROJECT LIFE CYCLE

Where in the life cycle of a project does planning begin? In the early conceptual phase, the main emphasis is on the results to be achieved by the project, with only a rough estimate of how the project will be carried out, how much it will cost, and when it can be accomplished. During each subsequent phase, more information is obtained, assumptions are gradually replaced or at least narrowed with known facts, and more detail is added to the project plan and schedule. In these early phases, relatively small amounts of money are usually spent ("seed money"), so detailed plans are not usually justified. However, at some point in all projects there comes a time when significant commitment of money and other resources is required. In a life cycle that is defined as having four phases—concept, definition (proposal), execution, and close-out—the point of significant commitment is generally at the start of the execution phase. A rule of thumb is that the definition phase will cost 10 times that of the concept phase, and the execution phase 10 times that of the definition or proposal phase. In many cases, these factors may be even larger. The close-out phase should expend only a fraction of the execution phase, if it is well-planned and controlled.

The material in this and subsequent chapters is directed toward the comprehensive planning and control required *prior to* approval of the significant commitment of resources at the start of the execution phase, often identified by acceptance of the project proposal. These planning and control principles and tools are equally useful in earlier phases.

The Importance of Adequate Project Planning

Inadequate project planning is a frequently cited reason for project failures. There are many causes of inadequate planning, including the widely recognized aversion of technical and, in fact, many other people to performing planning work ("Do you want me to do planning, or do some productive work?"), the reluctance of many people to expose their plans and knowledge of the job (or lack thereof) to others, and the basic difficulties of planning complex projects. In some situations, the complexities of imposed planning methods, techniques, and tools themselves block the creation of good plans.

In spite of these difficulties, creating a sound project plan is extremely important to project success. Without an adequate plan, the required resources cannot be assured and committed at the proper time, the team members cannot be fully committed to the project, monitoring and control will not be effective, and success will be a matter of luck.

Rules for Effective Project Execution

Thamhain and Wilemon,[1] in a study involving 304 project managers and their superiors from U.S. companies in electronic, chemical, and construction industries, identified a number of problems adversely affecting project performance and recommended several actions to assure effective project planning and execution:

- Assure that each team member is personally "signed-on" to the project.
- Work out a detailed project plan, involving all key personnel.
- Reach agreement on the plan among the project team members, the customer and sponsor.
- Obtain commitment from the project team members.
- Obtain commitment from management.
- Define measurable milestones.
- Attract and hold good people.
- Establish a controlling authority for each work package.
- Detect problems early.

Methods for building commitment in the project are presented in Chapter 6, some ideas for overcoming the barriers to integrated project planning are discussed in Chapter 7, and proven methods for project team planning are presented in Chapter 11.

10.3 PROJECT OBJECTIVES AND SCOPE

In Chapter 1, the hierarchy of objectives and strategies from the policy level to the strategic and operational levels is discussed. The specific objectives of each project must support one or more of these higher level objectives. It is important for the project sponsor, project manager, and functional project leaders to understand not only the project objectives but also the higher level objectives that the project supports. Only with this understanding can the best trade-off decisions be made when the inevitable conflicts between the time, cost, and technical results of the project occur.

Defining the Project Objectives

When a project enters the definition or execution phase, some statement of its objectives has normally been made. This is usually a description of the desired outcome of the project: *What* is proposed to be created; and a target date for its creation: *when* the results should be available. There usually will also be a statement of *how much* money can be spent in accomplishing the project. These three dimensions of the project objectives—results, time, and cost—form the core of the concrete, specific, or "hard" project objectives.

In addition, as discussed in Chapter 5, project objectives as criteria for project success can be viewed along a hard/soft dimension and an acceptable/excellent dimension. In addition to the classic "hard" time, cost, and technical performance objectives, the "soft" criteria are more subjective, and deal more with how the work is accomplished: attitudes, skills, behavior, and the more subtle expectations of the client. The acceptable/excellent dimension relates to the willingness of the project team to continually strive to exceed expectations. This does not mean that engineers should be encouraged to continually improve a design beyond the stated technical specifications (a practice that is frequently blamed for time and cost overruns on projects), but it does mean that the most successful project teams continually strive for improvement in all aspects of the project within the limits of time and cost that have been established. That includes improvement in the way things are done, to do them more quickly, efficiently, and with higher quality results, as well as improvement of the project results themselves. The key phrase is within *the limits that have been established.* Planning is about defining those limits, and determining how to achieve the project objectives within the limits.

The hierarchy of objectives and strategies described in Chapter 1 continues within each project. Given a good set of hard and soft project objectives, as elaborated by a project team using the team planning approach described in Chapter 11, the project manager and functional project leaders will each develop their own lower level strategies for achieving their pieces of the overall project objectives. These strategies will in turn have more detailed, shorter-term objectives, often tied to specific milestone events in the project master schedule.

Defining Project Scope

There is often confusion in project teams regarding the difference between the project objectives and the project scope. The term *project scope* refers to the "space or opportunity for unhampered motion, activity, or thought," "extent of treatment, activity, or influence," and "range of operation." A statement of project scope has been defined as "a documented description of the project as to its output, approach and content."[2] The terms *scope of work* or *statement of work* are often used, but they may or may not include the total definition of a particular project's scope. Project scope is given such importance in the project management literature that is recognized as one of the eight modules of the Project Management Body of Knowledge, as defined by the Project Management Institute (PMI), a professional society founded in 1969 with some 8,000 members worldwide. These modules are scope, time, cost, quality, risk, human resources, procurement, and communications.[3]

The statement of scope of a specific project must include

- The project results: What will be created, in terms of physical size and shape, geography, quantity, technical performance and operating specifications, cost characteristics, utility, and so on.
- The approach to be used: Technology (new or existing?), internal or external resources, definition of boundaries between the project and its environment.
- Content of the project: What is included and excluded in the work to be done, and definition of the boundary between the project tasks and other work that may be related to the project results or its environment.

Clear and complete definition of both the project objectives and the project scope are prerequisites for good project planning and control, and ultimate success of the project.

10.4   A PROJECT IS A PROCESS

It must be recognized that a project is not equal to the end results to be produced. The project is the *process* of creating a new result, which could be a new or modified product, facility, system, service, event (such as the Olympic Games every four years) or even organization. Unfortunately, most projects are named for their products, whether the product is a new hospital, a nuclear power plant, an orbital vehicle, a new consumer product, a new customer billing information system, or any of the many products created by projects. Thus, people look at the end result and think "That is the project." That is not true; the project is the entire process of creating the end result.

Every organization that plans and executes a project has—or must create—a process that the project flows through. Many organizations do not know or understand their development process, especially as an integrated, overall system. The specialized disciplines may each understand their pieces of the process, but rarely has anyone taken the time to document, analyze, and understand the overall picture, in other words, use the systems approach. As a result, the basic development process is often hidden, and broken into many bits and pieces.

Many project managers and their supporting planners simply re-invent the process, or what they think would be a good process, on every project. If they rely on the project team, as recommended and described in Chapter 11, the team members will each bring their understanding of their piece of the overall process to the party, and the team may capture a reasonable facsimile of the process as a result. But if the team had a good picture, preferably a flow chart supplemented as needed with written descriptions, of the organization's basic development process, they would be in a position to tailor that process to the needs of the specific project, and use the most efficient linkages of all parts of the process for their project plans.

Nicholson and Sieli, reflecting their experience at AT&T Bell Laboratories, propose that process management be integrated with project management. They advocate a process manager to be in charge of the overall product development process in the organization, in addition to project managers. "The integrating of project management with process management is a recommended approach for achieving continuous improvement in project management. The process manager is a resource to manage and improve the project management process, while the project manager focuses on the project."[4] These process managers report significant benefits from combining process and project management practices to the project managers they work with in projects to develop high-technology products.

10.5   THE PROJECT SUMMARY PLAN

A written project summary plan should be prepared by the project manager prior to or at the time that work on the project is started. This plan must explicitly define the objectives of the project, the approach to be taken, and the commitments being assumed by the manager.

For internally funded projects, the project summary plan should be approved before the work is started. For contractual projects the manager typically has up to 30 days to complete the plan in order not to lose valuable time during start-up.

The project summary plan should cover the following general topics:

- Project scope.
- Objectives (technical, profit, other).
- Approach (management, technical, make/buy).
- Contractual requirements (deliverable items).
- End item specifications to be met.
- Target schedules.
- Required resources.
- Major contributors (key team members).
- Financial limitations and possible problems.
- Risk areas (penalties, subcontractor default, work stoppages, technical exposures, etc.).

The plan must be complete, but not elaborate; precise, but not hairsplitting; thorough, but not constrained by a rigorous format.

As soon as the plan is complete, it is submitted to management for approval. When approved, it gives the project manager authority to execute and control the project to completion, following the general approach outlined in the plan, in conjunction with such documents as the contract or project release (described later), which authorizes the expenditure of money and other resources.

For contractual projects, the proposal may form a large part of the project summary plan. However, a number of very important parts of the summary plan are not included in proposals submitted to potential customers. Some companies require a complete project summary plan, accompanied by a written statement by the general manager that he has personally reviewed the proposed project and accepts the risks that it represents, before the proposal can be submitted to the potential customer.

For new product development projects, the product plans and research and development case application documents will usually cover the major topics of the project summary plan. The same is true for a well prepared project authorization request for capital facilities or new information systems.

Major changes in approach or other revisions to the plan made after initial approval must be communicated to management using the project evaluation and reporting practices described in Chapter 15.

10.6 PLANNING AND CONTROL FUNCTIONS AND TOOLS

This section lists the various planning and control functions and tools related to project management. No attempt is made to explain each of these many items in detail; rather, the intent here is to show how the numerous existing procedures and systems are interrelated to the several newer techniques developed to plan and control projects more effectively.

Product versus Project Planning and Control

It is beneficial to recognize the differences between the *product* (or results being produced by the project) and the *project* itself (or the process by which this product is being created). Figure 10.1 lists the major product planning and control functions and tools, and Figure 10.2 lists the functions and tools related to projects. These are presented to assure a comprehensive view of all the functions involved in effective project management. The project planning and control tools discussed more fully in this book are indicated by an asterisk in Figure 10.2.

Concern *what* will be the end results of the project.

Encompass technical specifications and drawings defining physical and performance characteristics of the product, and related procedures and practices.

Functions	Examples of the Tools
Defining, designing, and controlling the product characteristics	Product plans Market analyses Specifications, drawings, and diagrams Analytical techniques, tests, and reports Design review procedures Models, mock-ups and prototypes Cost estimating procedures Value engineering procedures
Defining and controlling product configuration	Drawing release procedures Design review procedures Configuration management practices Change order control procedures Change Control Board Production control systems Technical supervision
Establishing and controlling product quality	Quality control procedures Product assurance (reliability and maintainability) procedures Technical supervision Design review procedures Project evaluation meetings

Figure 10.1   Summary of product planning and control functions and tools.

10.7   PLANNING DURING THE CONCEPTUAL, PROPOSAL OR PRE-INVESTMENT PHASES

There are widely varying practices in the amount of planning performed before a project is authorized to begin. Inadequate planning during the proposal or pre-investment phase will result in funding and scheduling difficulties, and in some cases failure and cancellation of the project.

During preparation of the project proposal, considerable basic planning, estimating, and scheduling must be done to assure that the basic technical, cost, and schedule objectives to be proposed are attainable. A proposal team is formed, which ideally will be the nucleus of the future project team.

Concern *how* the end results of the project will be achieved.

Encompass administrative plans, systems, procedures and information defining the project objectives, work plans, budgets, resource plans, expenditures, and other management information.

Functions	Examples of the Tools
Defining and controlling project objectives and target completion date	Product plans DCF and other financial analysis methods Project authorization request (PAR) R&D case documents *The project file *Contract administration procedures *Project evaluation procedures *Risk analysis methods
Defining the deliverable end items and major tasks	Contract, R&D Case and/or PAR *Systematic planning techniques such as the *Project Breakdown Structure *Task/Responsibility Matrix *Master Schedule or Master Phasing Charts *Product development process definition *Project management process definition
Planning the work (tasks)	*Project Breakdown Structure *Work control package (task) definition procedures *Network systems (PERT/CPM/PDM) Bar charts and milestone charts *Risk analysis methods
Scheduling the work	*Network systems Bar charts and milestone charts Production control systems
Estimating required resources (manpower, money, material, facilities)	Working planning procedures Manpower and material estimating procedures Cost estimating and pricing procedures
Budgeting resources	*Planning, scheduling and estimating procedures *Budgeting procedures *Risk analysis methods
Work assignment and authorization: Internal	*Master Contract Release documents *Project Release documents *Work order documents
External	Contracting and purchasing procedures

*Discussed in some detail in the book.

Figure 10.2   Summary of project planning and control functions and tools.

Functions	Examples of the Tools
Evaluating progress: Physical	*Reporting procedures *Network systems (PERT/CPM/PDM) *Project management information systems
Cost	Financial information systems (accounting, budgeting, etc.) *Contract administration procedures *Project management information systems
Manpower	Manpower reporting procedures *Project management information systems
Schedule & Cost Control	*Budgeting procedures Work order procedures Contract change procedures *Project management information systems Production control systems
Technical	*Technical Performance Measurement procedures
Integrated evaluation of time, cost and technical performance.	*Project evaluation procedures and practices, using information produced by all the above tools

Figure 10.2   (*Continued*)

The ultimate project team organization is planned at this time, and the basic make or buy decisions are made for the major elements of the project. All of the planning steps described in the remainder of this chapter are taken, although in a fairly gross way. After the project is authorized to proceed, the planning steps are repeated and carried to the full degree of detail required to execute and control the project. The project summary plan is developed in outline form during the pre-investment phase, and prepared in detail after receipt of the contract or other approval to proceed. The foundation for success of the project must be firmly established during this critical proposal/planning phase.

During these early phases, the decision to proceed with the project must reflect the overall strategic management objectives, strategies, and decisions of top management, as discussed in Chapter 1. Project selection and justification are more closely related to strategic management of an organization than to project management per se. However, Morris[5] makes a strong case for "bring[ing] project management more strongly back into the project initiation stage of projects." Souder[6] describes a number of models (screening, evaluation, portfolio, organizational) for making the decision whether to proceed with a specific project or not. Pilcher[7] gives several basic, practical methods for economic analysis and risk appraisal of projects. In the conceptual or proposal phases, formal project management practices may not be applied to all projects, simply because so many of them fail to proceed to execution. However, application of project management principles in these early phases is a good investment of the time and money required, especially for high-risk projects that have a high probability of being approved for execution. During these early phases, risk analysis and risk management methods are of critical importance.

The Successive Principle: A New Logic for Planning under Uncertainty

During the conceptual phase of any project, and especially of a high-technology project, there will be many uncertainties concerning the project objectives, technologies, methods, schedules, and costs. The more concrete, quantified, and specific planning methods described in this book apply most appropriately to the later phases of a project when more specific information is available and the uncertainties have been reduced. When dealing with the rapid changes in today's world, managers have learned that the old, highly quantified methods of planning are no longer effective, and have adopted a planning logic that is more intuitive. Among the very few systems that are described in the literature, the successive principle is probably the most general and widespread. Lichtenberg[8] summarizes his experience in applying this principle to project planning, and identifies four basic conditions of the new planning logic:

1. Uncertainty is handled correctly and as a matter of greatest importance.
2. All matters of potential impact on the project are dealt with, including difficult and highly subjective issues and situations.
3. Only matters of relevance are dealt with.
4. The project is seen and planned as a whole, interlinked with its environment. Its various aspects (time, costs, resources, risks, etc.) are also recognized as being tightly interlinked.

Lichtenberg reports that feedback from managers who have been using the successive principle in their project planning for a number of years shows that this principle:

- Enables managers and decision makers to have a more realistic and better qualified preview of the potential end results of the project plans, even at a very early stage.
- Supports the teambuilding process and the mutual creativity of goals and objectives.

- Links planning, estimating, timing, risk management, resources, profitability, and environmental considerations into a harmonic whole.
- Works significantly faster and more flexibly than conventional planning procedures do.

Description of the successive principle. [9]   The Successive Principle is an integrated decision support methodology or process which can be used to address a variety of business problems or situations, and is particularly well suited to conceptualizing, planning, justifying, and executing projects. Its purpose is to produce unbiased, realistic results (time or cost estimates, risk analysis, profitability calculations, key decisions, and understanding of other key aspects or parameters of a project), based on holistic, broad coverage of all factors influencing or involved with the project, including subjective factors, hidden assumptions, and especially areas of uncertainty or potential change.

How it works.   The Successive Principle incorporates the concepts of holistic, whole-brain, systems thinking and the team approach with the mathematics of uncertainty and probability. The basic steps are:

1. *Identify the Evaluation Subject and Purpose:* The *subject* may be a set of strategic plans, a project in the embryonic or early conceptual phase, a response to a request for proposal or bid on a defined project, a project encountering unforeseen problems, a project entering a later phase of its life cycle, as well as other situations requiring a disciplined decision-making support system. The *evaluation purpose* may be to decide (1) whether to proceed with the project, prepare a proposal and submit a bid, (2) what action to take in response to a particular change or problem, and/or (3) what risks are involved and what contingency plans are required to mitigate the identified risks, to name a few examples.

2. *Form the Evaluation Team:* Identify the most appropriate team of people for the evaluation purpose. The team should include persons with knowledge and experience in the major aspects of the evaluation subject, and if possible representing the most involved organizations.

3. *Identify, Quantify and Rank the Central Factors of Uncertainty:* The team, using its factual knowledge and intuitive hunches and guestimates, stimulated by open interchange of ideas and opposing points of view in a truly collaborative style, first identifies the factors which, in their collective judgement, reflect the greatest uncertainties or unknowns regarding the evaluation subject and purpose. Frequently, this results in a list of "top twenty" items for further consideration. Second, the team members organize, define and

quantify each identified factor using the so-called triple estimate (minimum, likely, maximum) and Bayesian statistics to calculate the total result as well as the relative criticality to the result from each factor. This is expressed as the factor's specific influence upon the uncertainty of the result. (Correct use of the methodology eliminates the need for correlation factors because the user deliberately creates stochastic independence between items and factors.)

4. *Successively Break Down the Most Critical Factors to Reduce Uncertainty:* If any of the critical factors identified and quantified in Step 3 exhibit unacceptable levels of uncertainty (that is, the range between best and worst estimates is too great, or the mean value is too large or too small), the most critical factors are further broken down into their component parts (subsystems) by the team. These subfactors are in turn quantified and are included in the above ranking in the same manner as before. This successive breakdown, quantification, and ranking is continued until the level of uncertainty is close to the minimum or unavoidable. Logically, no further reduction in uncertainty can be achieved.

5. *Present the Results and Make the Decision:* The results of the evaluation are presented by the team to the decision maker, who may accept them or require a replanning. This systematic, disciplined, but wide-ranging and intuitive plus factual back-up for the results has proven to be extremely persuasive in many diverse settings, and the resulting decisions have proven to be well justified.

The results represent a realistic, largely unbiased total measure of the most likely values of the key parameters under consideration, and the related degree of uncertainty. Experience indicates that a list of the top ten areas of uncertainty will usually encompass all of the most critical items that need to be improved or kept under observation. Another very important but informal result from using the Successive Principle is attainment of a higher level of mutual understanding, trust and consensus among the evaluation team members. This improved potential for cooperation is utilized during execution of the project, if the decision is made to proceed, for better commitment, teamwork, motivation, and more productive response to unforeseen events and changes.

Case Example: Winning a High-Technology Systems Program Contract

Several typical applications of the Successive Principle are given here to illustrate its power and benefits. The first case involves a medium-large defense program, which might even be termed a mega-project (two-digit billions in US dollars), in a European country. The client country announced in the early 1980s its program for the radical modernization of a part of its defense system, and invited selected companies and joint ventures to a prequalification process. A company in a smaller country that was interested in proposing on this effort had a long experience and good competence in a part of the program, but suffered from a lack of know-how in other areas.

Initial use of the successive principle.   This company started a search for joint venture partners in other countries, and began negotiations with potential partners under the condition that the company would be the prime coordinator. At the same time the project manager, his senior management and key project persons performed their first analysis sessions using the Successive Principle to identify, clarify, and rank the business risks and opportunities imbedded in the venture as well as to estimate realistically the expected profit.

Useful results.   After the analysis sessions the participants found themselves changed. To a higher degree than before this group of managers and key project persons considered themselves as members of the same team. They also shared a deeper insight into the project, and found it to be most promising, but with a scope and complexity far beyond their previous projects. They conclusively established a firm team commitment to win the contract and then implement the project successfully. The project manager thus found himself having unusually strong support.

Negotiations with partners.   The knowledge of the primary problem areas obtained during the analysis was now used to support the company during its final negotiations with the joint venture partners, which soon resulted in agreements with companies in several countries.

Joint venture use of the successive principle.   After a successful prequalification of the proposed joint venture (JV), a JV project group of key persons was organized and a new series of analysis sessions was conducted using the Successive Principle. Guided by an updated "top-ten" list of uncertainty areas, the group strengthened itself as a team and at the same time worked systematically to clarify and utilize the largest potentials indicated in the top-ten list.

A key factor: The client country.   One of the top-ten factors was the degree to which the project was "colored" by and anchored in the client country. So the local share and local partners (those located in the client country) were given high priority and much attention at an early stage. In this connection it was found important to seek out not only competent, but also prestigious local partners. Because the joint venture came so early to this conclusion they could freely seek out and agree with the best of the local partners. This later proved to be a most important factor to final success.

Winning the final competition.   The next stage of the competition left three remaining candidates for the project: this joint venture and two competitors, both based in very large countries. These two competitors therefore were able to use considerable political pressure on their behalf, including visits of their countries' prime ministers. Parallel to the political side the client negotiated with the three candidates using the conventional forms of pressure. During this critical period, current analysis sessions (using the Successive Principle) allowed the project manager and his key persons to know exactly where he could give and where to hold fast, also keeping in mind the relative perceived benefits on each point as seen from the client's point of view. Most importantly, the joint venture knew, through the analysis results, the most realistic, ultimate limit of price reductions.

According to the project manager this knowledge, together with the top-ten lists, the team-building effect and the other benefits of the analysis sessions, were decisive factors for winning the contract, in spite of the handicap of being a considerably smaller company from a relatively small country.

The outcome to date.   The program has now been under implementation for some time. The updated expectations of the results are reported to be positive, and still corresponding to the earlier analysis results.

10.8   DEFINING THE PROJECT AND ITS SPECIFIC TASKS: THE PROJECT/WORK BREAKDOWN STRUCTURE

For many complex projects, it is necessary to use a disciplined, systematic approach to define the total project in such a way that all elements have the proper relationship to each other and that no element is overlooked. If this is properly done the result is very useful in a number of ways.

The most effective method of accomplishing such a definition of a project is the creation of a Project Breakdown Structure (PBS), also commonly called the Work Breakdown Structure (WBS). In the following paragraphs the PBS is described, its use is discussed, and several simplified examples are presented.

Description of the Project Breakdown Structure

The PBS is a graphic or word model of the project, exploding it in a level-by-level fashion down to the degree of detail needed for effective planning and control. It must include all deliverable end items (consumable goods, machinery, equipment, facilities, services, manuals, reports, and so on) and includes the major functional tasks that must be performed to conceive, design, create, fabricate, assemble, test, and deliver the end items.

Although the original impetus for defining projects systematically in this way came from large U.S. military and aerospace programs and projects, the concept has been adopted by virtually all areas of application of project management for complex projects. The basic U.S. Department of Defense definition is:

> *Work breakdown structure (WBS).* A work breakdown structure is a product-oriented family tree composed of hardware, services and data which result from project engineering efforts during the development and production of a defense/materiel item, and which completely defines the project/program. A WBS displays and defines the product(s) to be developed or produced and relates the elements of work to be accomplished to each other and to the end product.[10]

Because the term work breakdown structure is imbedded in the U.S. government project management directives and literature, this term has been used extensively in other industries. This name actually is somewhat misleading and causes confusion with practitioners, since it infers that only "work," or tasks, are involved. The tendency therefore is to look at a large project and immediately identify the work activities to be done—design, procurement, construction, commissioning, in a construction project. However, in most large projects, this is not a satisfactory second-level breakdown. Therefore the term *project* breakdown structure is used in this book, to emphasize that the many elements of the project must first be broken out, and then the functional work tasks identified for each deliverable element, at the proper level.

One Systematic Definition of a Project: The Project/Work Breakdown Structure

Prior to the advent of the project breakdown structure approach (which emerged in the early 1960s), the many different functional specialists who contributed to a project each broke the project down in different ways, to suit their particular needs. Many differing frameworks were developed, and still exist, for planning, estimating cost and other resources, budgeting, cost accounting, financial analysis, assigning responsibilities, purchasing, scheduling, issuing of contracts and subcontracts, material handling and storage, and many others. Often these frameworks are themselves different for the different functions of finance, marketing, engineering, procurement, manufacturing, field construction, and operations, to name a few. Bitter experience on many large projects showed that it was impossible to properly correlate and integrate the planning and control information on a large project in this hodgepodge of differing definitions of the same project. What was needed was one systematic project breakdown structure that all parties could agree to and understand for a given project, and to which all the other frameworks listed above could be cross-referenced. The PBS can be viewed as the Roseta Stone[11] of project management, as it enables the correlation of many diverse elements of information and coding schemes, like the diverse Egyptian languages which the Roseta Stone enabled scholars to understand.

Other methods of defining a project either break it down (1) according to the classic functions (design, engineering, manufacturing, production, construction, which often correspond to the basic organization structure) or (2) with a listing of contract line items. The advantage of the PBS over these methods stems from the hierarchical, structured approach, and the ability to visualize the total project and all its major and minor elements. The PBS breaks the project into a series of subprojects and deliverable items, then identifies the functional tasks needed to carry out the subprojects and create and deliver the deliverable items.

Creating a project breakdown structure. The project breakdown structure is developed by a judicious combination of the *product* breakdown structure with the *product development process* of the organization. The term product here refers to any result being created by the project, including hardware, software, equipment, facilities, services, staffed organizations, documents, data, and other tangible project results. The product development process, discussed earlier in this chapter, is the series of phases, steps, tasks, and activities that the organization uses in creating the various products of the project. Although this development process links the various phases and tasks together in a chronological flow, the PBS does not attempt to portray such chronological linkages. These will be identified in a subsequent planning step, when the project master schedule and action plans and schedules are prepared.

The PBS chart is created by starting at the top-level element, which identifies the total project, and breaking out the major, natural elements of the project (systems, facilities, categories of end items, and so on) at the next lower level. Each of these elements is then subdivided into its component elements. This level-by-level breakdown continues, reducing the scope, complexity, and cost of each element, until the proper practical level of end-item identification is reached. These are then subdivided into the major functional tasks that must be performed to create the end item in question. The objective is to identify elements and tasks that are in themselves manageable units clearly the responsibility of a functional project leader, and that can be planned, estimated, budgeted, scheduled, and controlled. This approach assures that the project is completely defined and that useful summaries of project information can be made.

During the creation of the PBS, which should be performed on a team basis as discussed in Chapter 11, the results or product breakdown plays a dominant role, but the product development process is also reflected at appropriate points. At the top level, the major phases of the development process are often identified, either directly or in terms of the major deliverables for a particular phase. For example, a major deliverable for the conceptual phase is often a *Conceptual Design Package* or a *Concept Feasibility Report.* Each of these might be broken down into sections or components, and then the functional tasks required to create each of those would be identified at the next lower level. The product development process heavily influences the definition of specific functional tasks or work packages at the lowest level of the PBS, and the organizational breakdown structure (OBS) will also influence the task definitions.

When first developing a PBS chart many persons tend to think of it as an organization chart. This results in a confused PBS overly influenced by the particular structure of the company involved. The PBS is *not* an organizational structure, although it may superficially resemble one depending on its drafting on paper. As discussed later, each element in the PBS can be identified as the responsibility of a particular person. One person may be responsible for a number of elements scattered throughout the PBS. The functional organization structure of the sponsoring company will, of course, have some influence on the PBS, especially at the level where deliverable end items are divided into functional tasks.

U.S. department of defense project/work breakdown structure guidance. Elaborate and detailed guidance has been issued officially by the U.S. Department of Defense (DOD) and the several individual DOD services for projects to develop and acquire their weapon systems, information systems, and other defense materiel items (see endnote 10). The basic standard provides general breakdowns for seven defense systems: aircraft, electronics, missile, ordnance, ship, space, and surface vehicle. Definitions are provided of the hardware and work breakdown through three levels for each system type. Work continues by a government/industry taskforce to reflect new technologies and more recent business practices, and a major revision of this standard is expected to be released in the early 1990s. A separate standard has been issued for information system and software projects.[12] The U.S. Department of Energy has also issued a guide for use in defining its major projects.[13]

PBS Examples

Figures 10.3 through 10.9 illustrate various ways of depicting a PBS for several types of projects. Actual breakdowns of large projects are usually

Level 1	2	3

Information System Development Project
    Concept Study
                    Feasibility Study
                    Requirements Study and Conceptual Design
                    Information System Proposal
                    Detail Design Specification
    Support Software
                    Compiler/Assembler
                    Data Base Control
                    Test Support Software
    Operating System
                    Executive
                    Input/Output Control
                    File Management
                    Message Handling
                    Control Management
                    Maintenance and Diagnostics
    Application Software
                    Application Program A
                    Application Program B
    Data
                    Technical Publications
                    Engineering Data
                    Support Data
    System Test and Evaluation
                    Equipment
                    Services
                    Facilities
    Training
                    Equipment
                    Services
                    Facilities
    Operations and Maintenance
                    Operations
                    Maintenance
    Project Support
                    Project Management
                    System Engineering

Figure 10.3   First three levels of a project breakdown structure for an information system development project. (Adapted from Postula, Frank D., "WBS Criteria for Effective Project Control," *1991 Transactions of the American Association of Cost Engineers,* AACE, Morgantown, WV, ISBN 0-930284-47-X, 1991, p. I.6.4.)

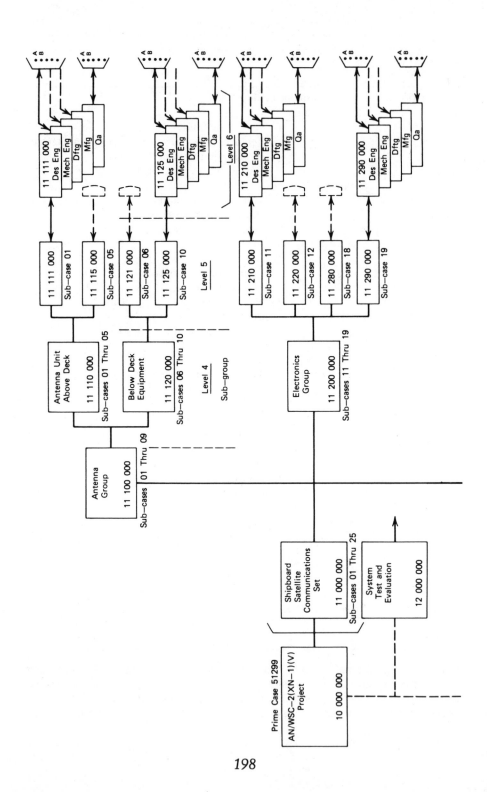

198

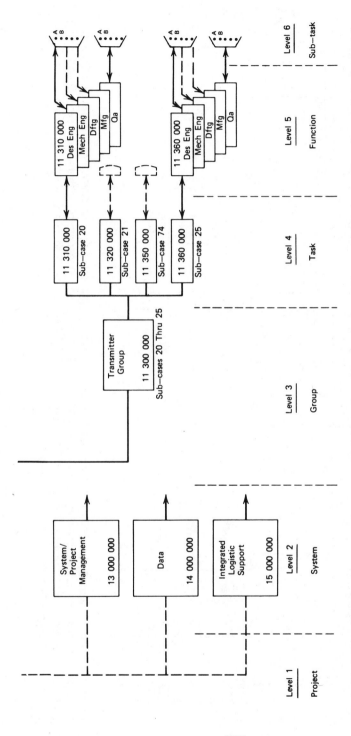

Figure 10.4 Breakdown structure for a large communications project (showing financial accounts assigned to project elements).

199

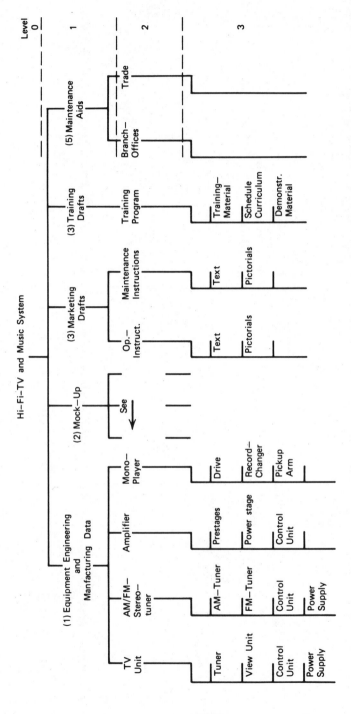

Figure 10.5 Sample format of project breakdown structure (PBS). RD and E portion of product development project: (1) RD and E has prime responsibility; (2) responsibility shared between RD and E Model shop; (3) responsibility shared between RD and E (first drafts of documents) and Marketing (final drafts); (4) responsibility shared between RD and E and Marketing Service. (Source: *ITT SEL Project Manager's Manual.* Used by permission.)

carried to four or five levels and often are presented on several pages to avoid excessively large charts.

Using the PBS

The process of creating the PBS produces, in itself, important benefits. In the process of breaking down the project, the project manager, the supporting planners, and key project team members are forced to think through all elements of the project. This helps to avoid omissions and clarifies the scope of work assigned to each functional project leader. The PBS is a means for visualizing the entire project in a meaningful way. Significant insight into the project and the interrelationships between its elements is often gained by creating and using the PBS. The following steps outline its customary use:

- Develop the initial PBS in a top-down fashion, through direct effort by the project manager and supporting planners, with appropriate inputs from the key project team members.
- Review and revise the completed PBS with all affected managers and team members and until agreement is reached on its validity.
- Identify work control packages (tasks) to be planned, estimated, budgeted, scheduled, and controlled.
- Identify for each PBS element down to and including each task:
    Responsible and performing organizations and functional project leaders.
    Product specifications.
    Prime and subcontracts and major purchase orders.
    Resource (people, funds, material, facilities, equipment) estimates and budgets.
    Work order numbers.
    Cost account numbers (task level only).
    Milestone events and related activities in PERT/CPM/PDM network plans, with scheduled dates.
- Summarize resource information up the PBS, comparing estimates, budgets, commitments, expenditures, and actual accomplishments for each element and for the project as a whole.
- Add expenditure to date to latest estimate to complete (ETC) for each task to obtain estimate at completion (EAC), and summarize up the PBS.
- Evaluate results to identify problems and initiate appropriate corrective action.

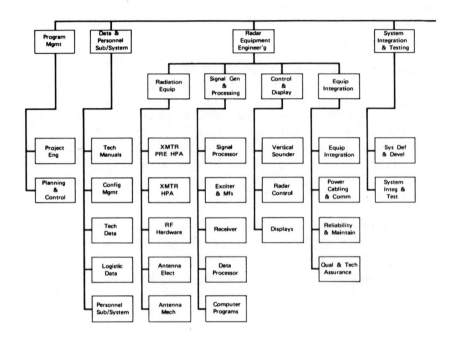

Figure 10.6   Program breakdown structure for and aerospace electronic system.

- Reiterate the above cycle as needed to replan and balance schedules, resources and scope of work.

The PBS thus becomes a framework of the project that enables all management information (from various systems and sources) to be correlated and summarized for planning and control purposes.

Controlling the PBS

When properly used, the PBS is a vital communication tool. It evolves and reflects current plans as the project matures. More detail is added to particular areas as the actual execution of those areas approaches. Thus a procedure for revising and controlling the distribution of the PBS is usually necessary to prevent unauthorized change and to assure that all managers are working with the latest issue. The PBS is, in a sense, the top assembly drawing of the project from the management point of view.

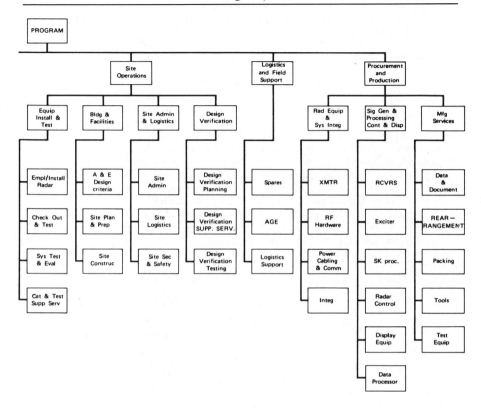

10.9   DEFINITION OF TASKS (WORK CONTROL PACKAGES)

Tasks are the final elements identified in the PBS, and thus are found at the end of the breakdown of a particular part of the project. Usually they emerge at several different levels of the PBS. These control packages may represent a total subcontract for work to be performed by an outside firm, but they usually represent a functional task for which a particular functional project leader is responsible. To be useful for control purposes, functional tasks should be of relatively short duration and small total cost compared to the total project duration and cost.

The cost accounting system is correlated to the PBS through the work control packages. A separate cost account number is established for each package or task, and expenditures are then recorded for each. The cost accounting coding scheme (chart of accounts) does not have to carry the

Level 0	Level 1	Level 2	Level 3	Level 4

Weapon system

Test instrumentation subsystem

Subsystem component RF set

Manufacture of common parts

Breadboard material

General task ODC (other direct costs)

This is part of the project breakdown structure for a very large electronic project. At the top level (level O) is the name of the total project, in this case a weapon system. In the above example, level 1 is for major subsystems. Level 2 elements are equipment groups for the subsystems. Level 3 items are breakdowns of these groups, and level 4 elements are support functions for level 3 tasks. Level 4 items here are the work control packages—elements small enough to control as entities.

Printed circuit boards

Excess material

Manufacturing tooling

Design of standard items

Program/project management

Product Assurance

Miscellaneous manufacturing

Miscellaneous project activity

Planning

Supply report

Reliability

Configuration

Power distribution center

    Manuals

    Design

    Manufacturing and support

    Test and support

Oscilloscope Station

    Manuals

    Design

    Design

    Manufacturing and support

    Test and support

Telemetry Control Communications

    Manuals

    Design

    Manufacturing and support

    Test and support

S-Band Signal Generator

    Manuals

    Design

    Manufacturing and support

    Test and support

Stack Assembly 01

    Manuals

    Design

    Manufacturing and support

    Test and support

(ETC)

Figure 10.7   Project breakdown structure for a very large electronic project.

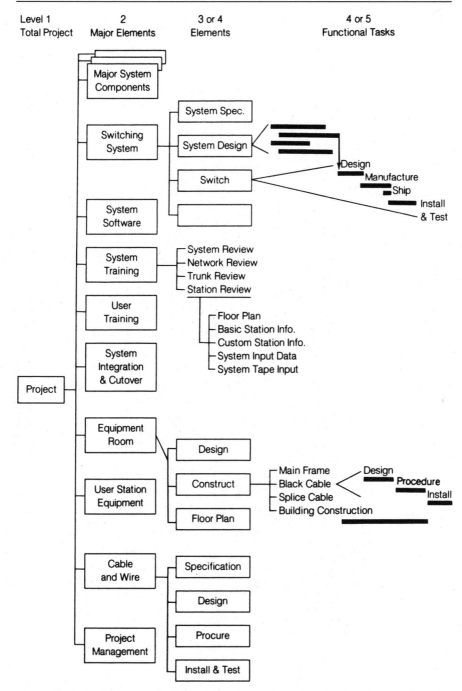

Figure 10.8   Project breakdown structure for a telecommunications-information system project.

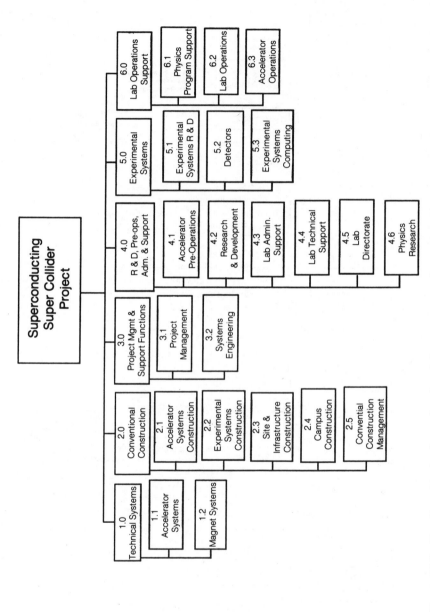

Figure 10.9 Top three levels of a project breakdown structure for a large, high-technology project. (Source: Superconducting Super Collider Laboratory, Universities Research Association. Used by permission.)

summary structure of the PBS. Summarization can be accomplished outside the cost accounting system for project management purposes, using the relationships established by the PBS.

Task Statement of Work

Each task is best defined by a concise statement of the work to be accomplished. This should include:

- Summary statement of the work to be accomplished.
- Inputs required from other tasks.
- Reference to applicable specifications, contractual conditions, or other documents.
- Specific results to be achieved: deliverable or intermediate items of hardware, software, documents, test results, drawings, specifications, and so on.

Types of Tasks and Effort

Several varieties of tasks can be identified, each requiring different treatment for scheduling and budgeting purposes. These include

- Design/development tasks.
- Manufacturing tasks.
- Construction or installation (field operations) tasks.
- Procurement (purchased and subcontracted) tasks.
- Management tasks.

Within these categories, three basic efforts exist as follows:

- *Readily identified tasks*. Specific start and end events associated with an end product or result.
- *Level of effort tasks*. Activities that cannot be associated directly with a definable end product or result, and that are controllable by time-phased budgets established for that purpose. Example: a project management task, including salaries and travel costs for the project manager and supporting staff when charged directly to the project.
- *Apportioned effort*. By itself not readily divisible into short-span tasks but related in direct proportion to other tasks. Example: inspection activity for manufacturing tasks.

Numbering Schemes for the Project Breakdown Structure

The numbering scheme applied to the PBS must be carefully designed to enable summarization of the schedule, cost, resource, and technical information from the task/work package levels to each intermediate level and up to the total project. Many organizations extend the numbering scheme to multiproject and corporate levels, to enable appropriate summarization across the entire organization, especially of resource information. The PBS numbering scheme should be kept separate from other corporate coding arrangements and not, for example, incorporated directly into the cost accounting numbering system. The PBS numbers must be compatible with the software capabilities of the project management information system, in terms of numbers of digits and summarization procedures. A typical PBS numbering system is illustrated in Figure 10.4.

Computer Software Support of the Project Breakdown Structure

An important feature of a project management computer software package is its ability to handle the PBS for scheduling, resources, and costs. Some of the more powerful packages allow the user to define the PBS on the screen, interactively and graphically, and provide for development of the project master schedule with milestones identified, intermediate level schedules, and task level schedules, all with appropriately integrated resource and cost information. This is an important capability to consider when selecting a project management software package, as discussed in more detail in Chapter 14.

10.10 TASK/RESPONSIBILITY MATRIX

The task/responsibility matrix is a planning tool for relating the work defined by the PBS to responsible organizational units, subcontractors, and individuals. With a given company organization breakdown structure (OBS) on one hand and the work to be performed as depicted by the PBS on the other, the objective is to couple the two. Since the PBS will never exactly match the OBS, the matrix provides a mechanism for assignment of prime and supporting accountability in the typical, functionally organized company.

Figures 10.10 and 10.11 illustrate this tool. Developing such a matrix will verify whether there is too much or too little detail in the PBS. It will also provide the planner with at least one activity on the PERT/CPM/PDM network plan for each entry.

When completed, the matrix provides a simple framework for further planning and control. In the planning process each prime task assignee

Project Breakdown Structure:	Project Team Members						
	Wade S. Proj. Mgr.	Bob B. Mktg.	Paul F. PSC	Larry H. FSO	Ken H. MMS	Tom L. Eng.	Etc.
Level 1 2 3 4 5							
New Telecommunications Project							
Electronic Switch System							
System Specifications —							
System Design —							
Switch Manufacture	I						
Place Orders	C	W					
Manufacture Switch Equipment	(Factory)						
Ship Switch Equipment	N			N		W	
Stage Switch Equipment On Site	N			N		W	
Switch Installation	I						
Make Area Ready	N			N		W	
Equipment On Site/Inventoried	N			W		W	
Install Switch	N			W		N	
Install Peripheral Equipment				W			
Test Remote Fibre Links				W			
(Etc)							

LEGEND

W—Does Work
C—Must Be Consulted
A—Must Approve
N—Must Be Notified/Copied
I—Integrative Responsibility

Figure 10.10   Example of a task/responsibility matrix.

should provide schedule, manpower, cost, and technical information on the assigned task, including the contributions required from those in supporting roles. The application of task and organization code numbers allows unique identification of each matrix entry. The logical and coherent framework of the matrix and associated technical information, schedules, and cost estimates provide the basis for work authorization and the necessary measurement criteria for control purposes. These points are discussed further in following sections.

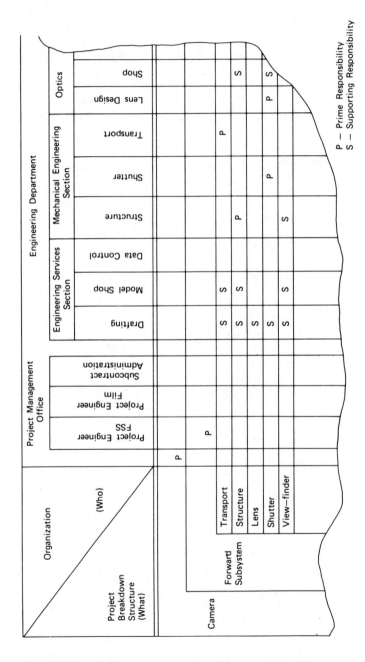

Figure 10.11   Task/responsibility matrix example.

210

The task/responsibility matrix is frequently referred to as a "linear responsibility chart (LRC)." Cleland and King[14] provide an excellent discussion of the use of this important tool in project management.

10.11 INTERFACE AND MILESTONE EVENT IDENTIFICATION

As an important part of the project planning effort, the major interface events to be managed by the project manager must be identified and incorporated into the detailed project plans and schedules. Milestone events are also important in developing the project master plan, evaluating overall progress, and reporting to higher management. Many, but not necessarily all, milestone events are also interface events.

Basic Definition of an Event

Poor project plans frequently result from improper or ambiguous event identification. An event is *not* a task or activity. *An event is an occurrence at a point in time that signifies the start or completion of one or more tasks or activities.* To be useful for project planning, scheduling, and control, an event must be understandable to all concerned, clearly and unambiguously described in precise terms, and its occurrence must be immediately recognizable. The occurrence of an event cannot be planned and scheduled to the hour, minute, and second of a specific day. But it must clearly be pinpointed to one specific day, month, and year; in other words, a calendar date.

Event Dates

Several dates may become associated with one event.

- *A scheduled date.* The currently agreed and committed time of event occurrence.
- *A predicted (earliest or forecast) date.* The currently predicted event occurrence time if the present plans are carried out without deviation or change.
- *A latest allowable (or required) date.* The time the event must occur, if the following tasks and events take place according to present plans, so that the project will be completed on schedule, or some intermediate contractual commitment will be met.
- *A target date.* The desired occurrence time.

- *A commitment date.* The time an event must occur, representing a customer or formal management commitment.
- *An actual date.* The time that the event actually did occur.

Not all of these dates are necessary for any given event, but confusion about the type of event date under discussion can cause serious communication problems.

Checklist for Interface and Milestone Event Identification

Figure 10.12 presents a checklist for key event identification. This is recommended for use in stimulating thought about interface or milestone events to be identified, planned, scheduled, and controlled.

To qualify as an interface event, an occurrence must denote a change of responsibility or a point of interaction between two or more elements of the project breakdown structure. A milestone event identifies a significant accomplishment in the project. Further discussion of interface events is given in Chapter 13.

10.12   THE PROJECT MASTER SCHEDULE AND THE SCHEDULE HIERARCHY

Schedule planning insures that all contract or other requirements, including hardware, software, and support items, are delivered on time. Fundamentally, there are two levels of schedule planning: the project level and the task level. The project level schedules integrate all tasks, interfaces, and milestones. Task level schedules and budgets are discussed later in this chapter. In very large projects intermediate levels of master schedules and budgets may be required, but these are considered to be extensions or subdivisions of the project schedules. (Although the word "task" is used here to denote a package of work that is the basic unit of project control, it is recognized that this word will have different meanings in various organizations. In adapting these practices, each organization should use the best term possible to avoid confusion and ambiguity.)

Types of Schedules

Many types of schedules are useful in managing projects, as indicated in Figure 10.13. The project manager and contributing functional project leaders should review this list to select those types of schedules required for each specific project. The various schedule types reflect emphasis on a specific element or function of the project, or combination of these. They are all derived from and consistent with the project master schedule.

1. *Things or objects moving through the interface or milestone events*

    Unique identification of

    General   Projects, systems, subsystems, project elements, requirements, funds.

    Documents   Contracts, subcontracts, specifications, drawings, plans budgets, schedules, charts, reports, work orders, procedures, bills of materials, parts lists, manuals.

    Equipment
    (Hardware) Models, test boards, components, materials, parts, subassemblies, assemblies, modules, deliverable hardware items, intermediate hardware items, equipment lots, spare parts.

    Software
    (Operating) Flow charts, coding lists, assemblies, packages, card decks, listings, tapes, source programs, object programs, compilers, simulators, systems, deliverable software items, intermediate software items.

    Services   Training, field support, operating, management, or administrative.

    Facilities Buildings or structures, test equipment or machines, tools and tooling, operating or production equipment, or machines.

2. *Operations performed on the things or objects*

    Begin, start, establish, define, analyze, design, modify, release, issue, procure, purchase, fabricate, assemble, wire, test, pack, ship, receive, inspect, hang, mount, place, install, adjust, accept, approve, operate, negotiate, write, prepare, compile, correct, collect, construct, complete, finish, end.

3. *Event designator*

    Past tense of any verb in Section 2 associated with one or more objects in Section 1.
    Examples: "Contract XY funds released"
                "Hardware lot 2 received on site"
                "Tests complete on software package no. 3"
                "Building complete and available"

Figure 10.12   Checklist for interface and milestone event identification.

The Project Master Schedule

The project master schedule interrelates all elements and tasks of the project on a common time scale. It should

- Be based on the PBS.
- Be complete and comprehensive in scope.
- Reflect contractual commitments and customer obligations.

1. Project master schedule (Master Phasing Schedule)
2. Major milestone schedule
3. Master development schedule
4. Master production schedule
5. Master summary schedules—PBS level 2, 3, etc.
6. Near term milestone schedule
7. Project trend charts (cost/milestones).
8. Project network plan (PERT) (management logic diagram)
9. Task schedules
10. Functional element schedules
11. Detailed PERT network(s)
12. Major product schedules
13. Hardware utilization schedules
14. Unit hardware delivery schedules
15. Preaward subcontract schedules
16. Subcontractor submitted engineering, manufacturing, and procurement schedules/PERT.
17. Drawing release schedule
18. Contractually required review schedules
19. Flight test schedule
20. Government or customer furnished property schedule
21. Government or customer data or other obligation schedule
22. Training equipment and maintenance demonstration hardware schedule
23. Technical publication schedule
24. Material support test schedule
25. Contract data requirements schedule
26. System demonstration schedules

Figure 10.13   Types of schedules used on projects.

- Assist in planning the buildup and effective use of manpower and other resources.
- Include key interface and milestone events linking all tasks.
- Be useful for progress evaluation and management reporting.

The project master schedule is initially established during the proposal (or equivalent) phase. It is continually refined as the project progresses. It is developed in an iterative top-down, bottom-up fashion by the project manager and supporting planners, working with the concerned functional project leaders as the task schedules and budgets are developed.

A *management schedule reserve* should be established at the outset of the project by setting the planned completion date ahead of the critical commitment date by some reasonable amount of time. This provides a contingency reserve to be carefully allocated by the project manager when specific tasks encounter unavoidable delays that affect the project critical path. This reserve is analogous to the management budget reserve discussed later.

The project master schedule may take the form of a bar chart, a bar chart with selected interface and milestone events, or a time-scaled summary PERT network or logic diagram. If the last, it will usually appear as a bar chart or "bar-net" with selected interface and milestone events, and with logical dependencies shown between the events. Where intermediate level master schedules are required, different formats may be needed for different schedules, as indicated. Figures 10.14, 10.15, 10.16 and 10.17 illustrate several types of master schedule formats.

Computer-Generated Master Schedules

Many of the better project scheduling computer software packages will produce high quality, multicolor project master schedules on plotters and inkjet printers, or black and white schedules on laser and dot matrix printers. The best of these also portray milestones and interface events, with user control of the symbols used. One example of a computer-generated master schedule is shown in Figure 10.17.

10.13   THE PERT/CPM/PDM PROJECT LEVEL NETWORK PLAN

Application of the PERT/CPM/PDM network planning and critical path analysis technique to the overall project can produce significant benefits, if done properly. These benefits include:

- Integration of all tasks with interface and milestone events.
- Reduction in total project duration through improved overlapping of tasks and activities, where feasible and necessary.
- Identification of the chain of events and activities leading to project completion, which forms the *critical path*. These are the activities and events that, if delayed, will delay the project completion; and that, if accelerated, will enable earlier project completion.
- More effective integrated evaluation of actual progress by all contributors.

The basic elements of a network plan are:

- *Events.* The start or completion of one or more activities or tasks; represented graphically by a circle, square, or other geometric figure.
- *Tasks or activities.* Time-consuming jobs or actions; represented graphically in one of two ways: (1) in CPM (critical path method) and PERT (Program Evaluation and Review Technique), as lines or arrows between events, with the arrowhead showing the sequential dependencies

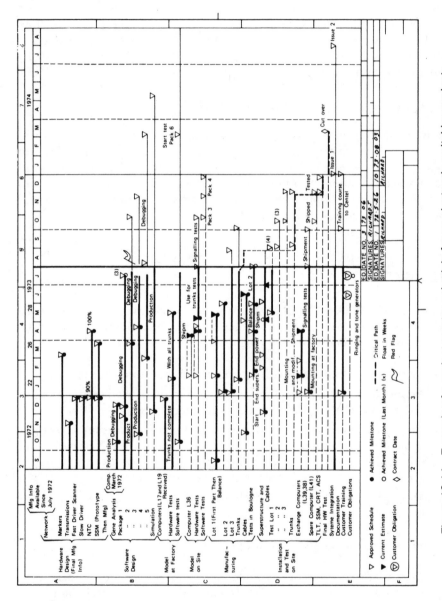

Figure 10.14 Example of a project master schedule for an electronic switching project.

216

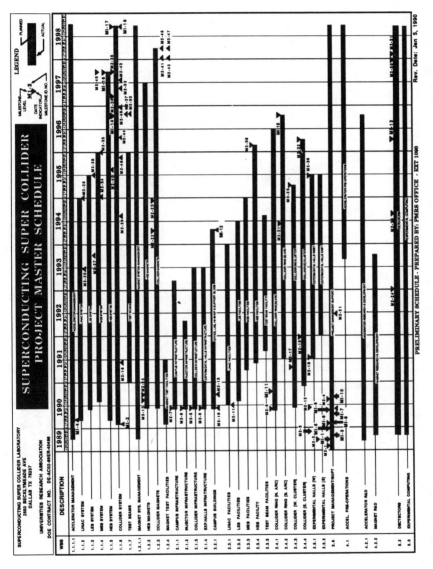

Figure 10.15 The project master schedule for a large, high-technology project. (Source: Superconducting Super Collider Laboratory, Universities Research Association. Used by permission.)

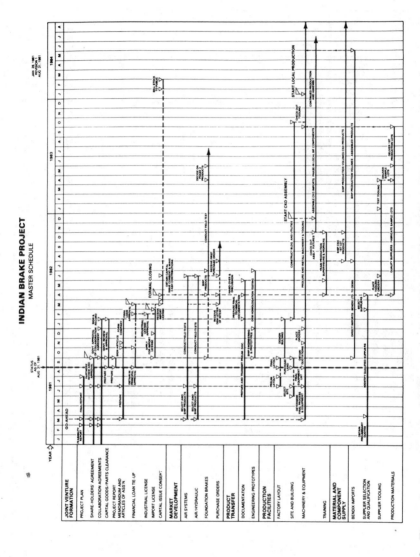

Figure 10.16 The project master schedule for a complex joint venture project involving a new company, technology transfer, a new manufacturing plant, new suppliers, and new customers.

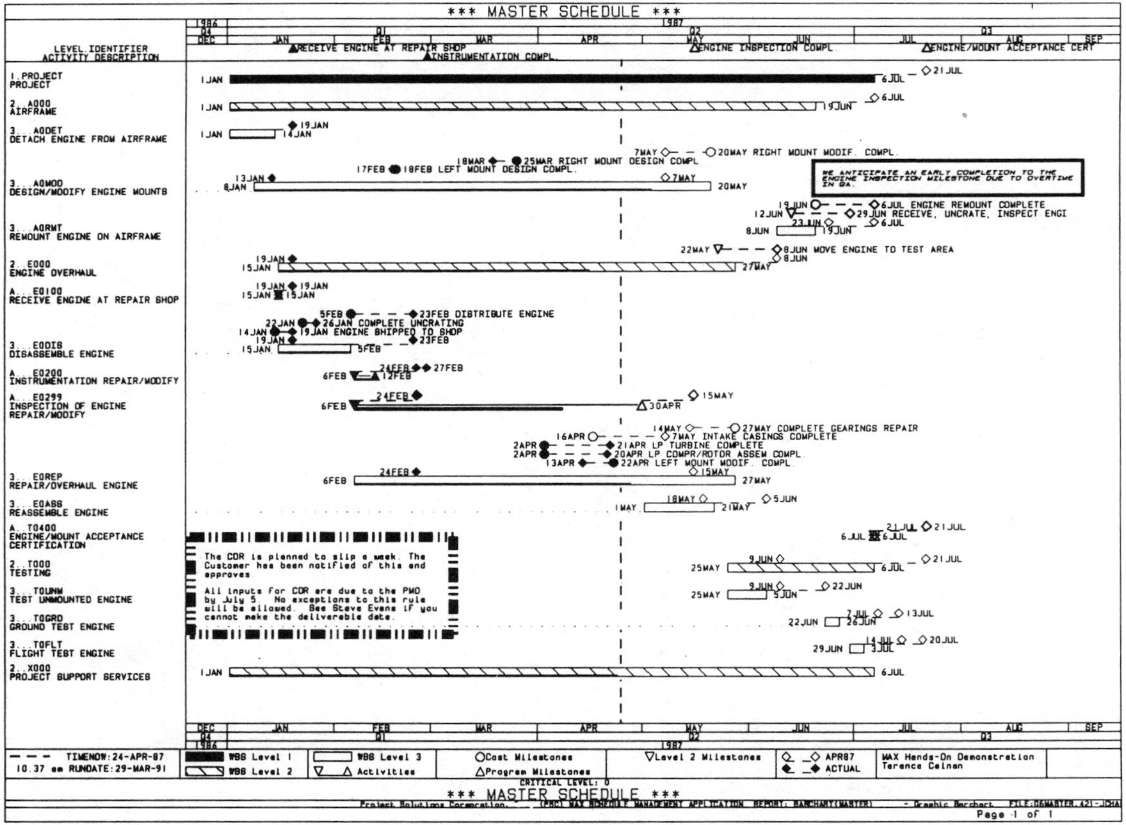

Figure 10.17 Computer generated level 2 schedule (partial). (Source: Project Solutions Corporation. Used by permission.)

219

involved; and (2) in PDM (precedence diagram method), as enclosed geometric shapes, usually rectangles. In CPM plans, the events are usually small circles with numbers only; in PERT plans, the events are described and coded. Hence, CPM plans are often termed "activity oriented" and PERT plans "event oriented."

- *Dependencies*. The logical constraints between the events and activities. On CPM and PERT network plans, the dependencies are indicated by the graphic connections made between the events and activities. On PDM plans, arrows are drawn between the activity boxes to indicate the dependency relationships. Dependencies in CPM and PERT plans are simple: when the preceding activity is 100 percent complete its end event has occurred and the succeeding activity can start. In PDM plans, more complex dependency relationships can be shown: finish-to-start (as in CPM and PERT notation), finish-to-finish, start-to-finish, and start-to-start, and these can have lead and lag times associated with them. PDM plans generally require fewer activity boxes than CPM or PERT plans to portray a complex project plan.

Figure 10.16 illustrates the typical appearance of portions of a network plan when plotted on a fixed time scale. Figures in later chapters illustrate project and task network plans and schedules.

Time Analysis of Network Plans

After the tasks or activities in a project have been defined and the network plan developed showing the logical sequence of the work, the durations and required resources for each task or activity are estimated. Then by simply adding up the times from left to right (the "forward pass") the time required to complete the project *according to this plan* can be determined, together with the "expected" start and completion dates for each activity and event. Next, by starting with a given date at the end of the network plan and subtracting the durations, working backward through the plan (the "backward pass"), the "latest allowable" dates for each event and activity can be calculated. The *critical path* consists of the chain(s) of activities whose expected and latest allowable dates are equal; these have zero "float" or "slack" time (delay allowable without delaying project completion). Since on average about 15% of the activities in a project are on the critical path, all others have some float or slack time, which can be used to balance the resource requirements.

There is a large quantity of published literature concerning PERT/CPM/PDM network planning and critical path analysis and the application of the technique to projects in various industries. Mulvaney's[15] short book on preparing and using PDM network planning provides a complete description of this powerful technique, and O'Brien[16] provides an in-depth treatment of its use in the construction industry.

Effective application of this technique for project planning and control can be difficult. In brief, the following principles should be observed:

- Maintain the master schedule emphasis in the project network plan.

    Incorporate all interface and milestone events.

    Organize the network to reflect the PBS.

    Assure that all specific tasks are represented, but exclude level-of-effort and apportioned tasks.

    Avoid detail more appropriate to the task schedules (see Section 10.15) or short interval schedules (production control).

    Include customer and other external obligations and constraints.

- Use the project network plan to validate, substantiate, and determine ways to recover to the master project schedule; change the master schedule only by appropriate management decision, not automatically to reflect the current network plan.

- When it is necessary to use electronic data processing (EDP) support, choose the software package carefully; inflexible or inadequate software can be the source of extreme aggravation, delays, and extra costs (see Chapter 14).

- Display the results produced by the analysis of the project network plan on a fixed time scale for higher management review; use EDP plotting equipment where economical and feasible.

- Conduct adequate indoctrination of all concerned to assure understanding and use of the results.

Project network plans are usually rather large charts and vary considerably in method of documentation, hence it is not feasible to reproduce examples in this book. Further discussion of the use of PERT/CPM network plans at detailed task levels is given in Section 10.15.

10.14 THE PROJECT BUDGET AND RESOURCE PLANS

The resources to be planned, budgeted, and expended or otherwise used to carry out the project include: time, money, people, facilities, equipment, and materials.

The nature of each specific project dictates which resources are critical to the project and therefore must be scheduled carefully. Time is budgeted on a gross basis by the project master schedule, and in detail by the task schedules. Facilities, equipment, and material are planned, budgeted, and controlled by special procedures set up for those purposes within each organization. The project master schedule establishes the key dates to be used in those procedures, and in setting up the detailed task schedules. The resources of money and people require special attention for effective project planning and control.

The *project budget* is equivalent to the operating budget for an organizational unit. A key difference is that, rather than being on an annual recurring basis, the project budget covers the entire project through to its completion. The project budget should be divided into direct and indirect project budgets for effective control.

The *direct project budget* is a primary control tool for the project manager and the contributing functional managers. It includes the costs (labor, travel, etc.) incurred by all project team members performing specific project tasks, as well as:

Cost of resale material

Standard manufacturing cost

Manufacturing variances

Engineering expense

Installation (field operations) expense

Freight and delivery

Other direct cost

An example of a Direct Project Budget summary is given in Figure 11.2, Chapter 11. The direct project budget should:

- Be broken down following the PBS exactly to the task (work control package) level.
- Include for each task and intermediate level project element the labor, material (purchased or subcontracted), and other costs as discussed later under task schedules and budgets, on a weekly basis.
- Include overhead or burden on direct labor or material.
- Be summarized also by contributing or performing organization, cutting across the various tasks.
- Provide management reserves, as discussed in following paragraphs.

The *indirect project budget* includes warranty or penalty costs; research and development assessments; service charges and commissions; financing, marketing, and general administrative costs; inventory adjustments; and other allocations. These are generally set up for the total project without further breakdown. They should be extended by quarter through the life of the project to assist in preparation of the project funding plan, if required.

The *total project budget*, combining both direct and indirect budgets, includes the *gross and net profit projections.*

The project budget is established during the proposal (or equivalent) phase. Initially it is a gross target. As the project breakdown is developed, more detailed estimates are made at the task level by the functional managers responsible for performing the work. Through reiterative top-down, bottom-up cycles, the project and functional managers negotiate mutually agreeable cost estimates and budgets for each task. When the project is authorized to proceed (contract award, or approval of research and development case or project authorization request), further negotiations reflect any changes from the proposal estimates. The project budget, supported by the individual task budgets, is then established and approved, and the work is authorized to proceed as described in Chapter 11. It is recommended that existing budgeting formats be adapted within each unit to serve its project management needs.

Management Reserves

It is a natural tendency on the part of every person to provide a certain amount of cushion or reserve in his or her time and cost estimates. If this tendency is not strictly controlled, overall project cost estimates become substantially inflated. If approved, then each person will tend to expend all the time and cost available to him, including his reserve for contingencies. As Parkinson[17] has said:

- The work at hand expands to fill the time available.
- Expenditures rise to meet budget.

The project manager must strive to identify all such reserves for contingencies, both in time and cost. As a general rule, these should be collected and held at the highest possible aggregate level in the project. Each task should be scheduled and estimated realistically. It should be recognized that only a few tasks should encounter major unforeseen difficulties. The management problem is that it is not possible to predict which tasks will have such problems. Therefore, by establishing a central management reserve, the comptroller and project manager have the freedom to allocate funds from the reserve to overcome the unforeseen problems. This will minimize rebudgeting funds for all other tasks. If the reserve is not totally used, the balance is then available as added profit.

For these reasons, it is recommended that only one contingency reserve account be established on each project, with appropriate procedures for allocation to various tasks as needed during execution of the project. This requires setting tight but realistic targets for every task. A figure of 10%

of direct costs is commonly used for the initial contingency on complex projects.

If unrealistically low initial budgets are established or large cost increases are experienced, the contingency reserve may be negative. The project manager must then seek additional funds from his management, while attempting to find places to reduce the scope or quality of the work to reduce the task budgets to the level required.

A *management reserve transaction register* is necessary to assure adequate control of the reserve. Appropriate procedures for approval and control of reserve transactions must be established, with the project manager as one of the approving authorities. These principles should be observed:

- Reserve transactions do not affect the original task budgets, which should remain fixed.
- *All* transfers between tasks should go through the reserve transaction register, including transfers from tasks where underruns are experienced, as well as transfers to tasks requiring additional funding.
- Specific approvals should be established for:

  Contractually authorized increase or decrease in scope of work

  Cost savings realized, either during or after task completion

  Cost expenditures over budget.

Project Funding Plan

It is the responsibility of the company comptroller to assure the adequacy and validity of funding proposals. However, in some cases, the project manager will be responsible for preparing and obtaining approval of a *project funding plan*. This is based on a cash flow analysis of the project budget coupled with the project master schedule, to identify expenditure levels by month and thereby establish the funding requirements. The project manager would then identify the sources of funds for the life of the project, based on contract payment terms, related research and development cases and PARs (project authorization requests), with assistance and approval of the comptroller. The Project Funding Plan would thus indicate the working or invested capital required to be provided by the unit each month or quarter in support of the project. The funding plan should include termination costs, if any, over and above the project budget, should the project be terminated before completion.

Project Chart of Accounts

During the development of the project master schedule and budget, the project chart of accounts should be established. This is necessary to assure proper work authorization, accounting for expenditures, and cost control, as discussed later.

By analyzing the task/responsibility matrix (Figures 10.10 and 10.11), the project manager, project accountant, contract administrator, and project controller, working with the functional managers as appropriate, can establish the project chart of accounts so that all accounting and reporting requirements can be satisfied. Important considerations here include the following:

- Each task (or work package) that is to be scheduled, budgeted, and controlled should be assigned a unique cost account number within the corporate chart of accounts.
- Summaries by elements of the project breakdown structure do not *necessarily* have to be produced by the accounting system; in other words, the project chart of accounts and the accounting numbers assigned do not have to reflect the PBS, although this is desirable if possible.
- Summaries by PBS elements can be prepared manually or by simple EDP programs outside the basic cost accounting system, but using data produced and controlled by the accounting system.
- A reasonable balance must be maintained between the improved control that a large number of cost accounts provides and the added cost and administrative burden resulting from more numerous accounts.

Further discussion of this area is given in Chapter 12.

10.15   TASK PLANS, SCHEDULES, AND BUDGETS

Task Schedules

A task schedule covers all key activities that must be carried out to complete the task as defined in the task statement of work. It should

- Be prepared by the responsible performing manager or functional project leader and negotiated with the project manager.
- Incorporate all related interface and milestone events from the project master schedule.
- Show a reasonable number of specific activities that can be related to cost, manpower, material, or other resource requirements.
- Enable monitoring and evaluation of progress on the task.
- Reflect not only the project schedule requirements, but also the functional organizational requirements and resources considering all other tasks in the organization.

Task Budgets

The budget for each task is based on the cost and resource estimates to accomplish the work, together with the task schedule. In practice, the schedule and budget are usually developed simultaneously with the objective of meeting the limitations imposed by the project master schedule and budget. The task budget should

- Be prepared by the responsible performing manager, using available estimating and pricing procedures and specialists, in negotiation with the project manager.
- Be related directly to the activities shown in the task schedule, to the extent possible.
- Enable collection of and comparison to actual expenditures of money, manpower, and other critical resources.

Design/Development Task Schedules and Budgets

The design/development category includes tasks involving purchasing, engineering, drafting, model shop, product engineering, engineering test, prototype fabrication and assembly, and similar functions as related to design and development of the end product or services.

A suggested design/development task schedule and budget format for labor hours, and material and other direct costs is shown in Figure 10.18. Various detailed estimating forms and procedures are ordinarily used to assist in completing the design/development task schedules and budgets. Labor cost and labor and material overhead cost should be added to the information given in Figure 10.18 to produce a complete task budget.

Manufacturing Task Schedules and Budgets

Based on the product design as specified in the proposal, contract, research development case (if any), and available engineering documentation, Manufacturing prepares the cost estimates and budgets for each manufacturing task. The production control group establishes the manufacturing delivery schedules. Both of these actions are accomplished in a negotiating process between the manufacturing project leader and the project manager, with appropriate involvement of the project engineer and other project team members. Both the budgets and schedules must be consistent with the project master schedule and budget. Variances in time or cost must immediately be referred to the project manager or higher line authority for resolution.

Manufacturing tasks may involve either recurring or nonrecurring costs, or both. Recurring costs are those associated with normal production

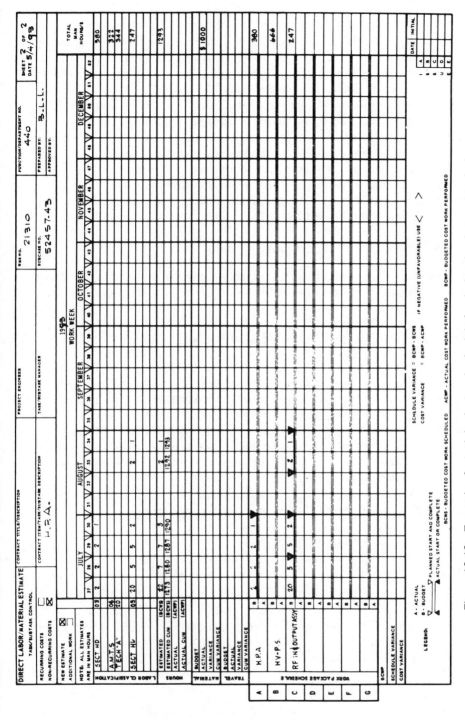

Figure 10.18 Example of a task budget and schedule for design development tasks.

227

operations for specific tasks, as indicated in Figure 10.19. Nonrecurring costs are associated with tooling and special test equipment, and similar special items, as well as industrial, fabrication, test, packing, and product engineering for a specific task, as indicated in Figure 10.20.

Other established manufacturing estimating procedures and forms are used to develop the labor and material costs for each manufacturing task in the project. Where standard production costs are available for manufactured parts or assemblies to be used on the project, detailed estimates such as shown in Figures 10.19 and 10.20 would not be required.

Manufacturing task schedules will generally show deliveries of the parts or assemblies produced by the task on a quantity per week basis by part number. These should be correlated to budgeted costs for control purposes.

Installation (Field Operations) Tasks

Installation or field operations tasks may be scheduled and budgeted using formats similar to those applicable to design/development tasks (Figure 10.18).

Procurement Task Schedules and Budgets

For complex tasks being performed under subcontract or purchase order by outside organizations, detailed schedules and budgets are required. These may be for design/development, manufacturing, or installation (field operations) tasks, and the requirements stated in the preceding paragraphs would apply. Usually, however, less detailed schedules and budgets will be provided by subcontractors and vendors. If detailed schedules are required, the contract or purchase order should clearly stipulate such requirements.

Budgeted and Actual Cost for Work Scheduled and Performed

Budgeted cost for work scheduled (BCWS) is the budgeted amount of cost for work scheduled to be accomplished, plus the amount of level of effort or apportioned effort scheduled to be accomplished, within a given time period.

Budgeted cost for work performed (BCWP) is the budgeted amount of cost for completed work, plus budgets for level of effort or apportioned effort activity completed within a given time period. This is sometimes referred to as the "earned value."

Actual cost for work performed (ACWP) is the amount reported as actually expended in completing the work accomplished within a given time period.

MANUFACTURING FUNCTIONAL COST—TASK BUDGET

PROPOSAL TITLE		PROPOSAL NO.	DATE
CUSTOMER			
ITEM/TASK DESCRIPTION		ITEM NO./TASK NO.	QUANTITY

MATERIAL INCLUDING SUB-CONTRACTING

PURCHASED PARTS	UNIT COST	TOTAL COST	FREIGHT & SHRINK. @ _____ %	TOTAL INCLUDING FREIGHT & SHRINKAGE
PACKING				
SUBCONTRACTING				

REMARKS		
	TOTAL W/O OPERATION BURDEN	
	OPERATION BURDEN @ _____ %	
	TOTAL LOADED MATERIAL	
	REVIEWED AND APPROVED BY AND DATE	

DIRECT LABOR

NO.	FUNCTION	LABOR CODE	UNIT HOURS	TOTAL HOURS	RATE	DOLLARS		DOLLARS
	PROGRAM MANAGEMENT						DIRECT LABOR W/O ECONOMIC FACTOR	
70	FABRICATION MACHINIST	41					ECONOMIC FACTOR @ _____ %	
70	FAB. INSPECTION ____ %	71						
60	ASSEMBLY "A"	51					TOTAL DIRECT LABOR	
60	ASSEMBLY "B"	52					OVERHEAD @ _____ %	
53	ASSY. INSPECTION ____ %	71					TOTAL LOADED LABOR W/O OPERATION BURDEN	
71	PACKING AND SHIPPING	55					OPERATION BURDEN @ _____ %	
30	INDUSTRIAL ENGR'G.	61					TOTAL LOADED LABOR	
20	PROD. CONTROL PLANNER	62					TOTAL LOADED MATERIAL	
52	TEST	72					TOTAL FACTORY COST	
							UNIT FACTORY COST	
			TOTALS					

REMARKS	
	PREPARED BY DATE
	CHECKED BY AND DATE
	REVIEWED AND APPROVED BY AND DATE

RECURRING COSTS	ITEM NO./TASK NO.	QUANTITY

MK5064 12/67

Figure 10.19   Manufacturing functional cost- task budget for recurring costs.

	MANUFACTURING FUNCTIONAL COST–TASK BUDGET		
	PROPOSAL TITLE	PROPOSAL NO.	DATE
	CUSTOMER		
	ITEM/TASK DESCRIPTION	ITEM NO./TASK NO.	QUANTITY

MATERIAL INCLUDING SUB-CONTRACTING

TOOLING AND TEST EQUIPMENT	UNIT COST	TOTAL COST	FREIGHT & SHRINK. @ _____ %	TOTAL INCLUDING FREIGHT & SHRINKAGE
TOOLING (VENDOR)				

OTHER

TOOLING (ASSEMBLY)				
TOOLING (FABRICATION)				
SPECIAL TEST EQUIPMENT				

REMARKS		
	TOTAL	
	OPERATION BURDEN @ _____ %	
	TOTAL LOADED MATERIAL	
	REVIEWED AND APPROVED BY AND DATE	

DIRECT LABOR

NO.	FUNCTION	LABOR CODE	TOTAL HOURS	RATE	DOLLARS		DOLLARS
30	INDUSTRIAL ENGINEERING	61				DIRECT LABOR W/O ECONOMIC FACTOR	
70	FABRICATION ENGINEERING	61				ECONOMIC FACTOR @ _____ %	
51	TEST ENGINEERING	72				TOTAL DIRECTOR LABOR	
71	PACKING ENGINEERING					OVERHEAD @ _____ %	
40	PRODUCT ENGINEERING					TOTAL LOADED LABOR W/O OPER. BURDEN	
						OPERATION BURDEN @ _____ %	
						TOTAL LOADED LABOR	
						TOTAL LOADED MATERIAL	
						TOTAL FACTORY COST	
	TOTALS						

REMARKS/ADDITIONAL CAPITAL EQUIPMENT	PREPARED BY AND DATE
	CHECKED BY AND DATE
	REVIEWED AND APPROVED BY AND DATE

NON-RECURRING COSTS	ITEM NO./TASK NO.	QUANTITY

MK5065 ISS. 12/67

Figure 10.20   Manufacturing functional cost- task budget for nonrecurring costs.

These costs apply to individual tasks or work packages (completed, in-progress, or future, as appropriate), to summaries of several tasks within a higher level element in the project budget structure, or to the total project. They are identified in Figures 10.18 and 15.2.

These values are useful in calculating schedule and cost variances in dollars or other monetary units at each level of summarization on the project (task, intermediate elements, and total project)

$$\text{Schedule variance} = \text{BCWP} - \text{BCWS}$$
$$\text{Cost variance} = \text{BCWP} - \text{ACWP}$$

The use of these values for control and evaluation purposes is discussed further in Chapter 12.

10.16 INTEGRATED, DETAILED TASK LEVEL PERT/CPM/PDM PROJECT NETWORK PLAN AND SCHEDULE

For some projects considerable improvement in overall scheduling and control will result if all appropriate tasks are planned in network fashion, and all such task networks are integrated by placing them into the appropriate parts of the project master network. Level of effort and apportioned activities would be excluded from the detailed, integrated network.

Although this approach is desirable in general, careful planning and control of the planning effort is required to avoid or overcome a number of potentially serious difficulties.

- If the task networks are too detailed, the resulting integrated project network becomes costly and difficult to maintain on a current basis.
- Electronic data processing procedures are usually required; useful, flexible software must be available, and quick response processing of results must be provided.
- Thorough indoctrination and training must be given to all task leaders and other users.
- The project control group must realize they are serving the functional task managers and leaders in providing them with the scheduling information produced, and not controlling the internal schedules of each task.
- The "rolling wave" technique (detailing plans only for the next few months) is a useful way to avoid excessive detail.

Despite many years of experience in the application of PERT/CPM/PDM network planning through many industries, its successful use at the detailed, integrated level for large projects is difficult to achieve. It is recommended that such application be limited to the project master schedule level until considerable experience is gained with its use. Then, selected critical tasks should be added on a detailed basis. Experience on successive projects will indicate the practical limits within particular environments and for specific types of projects.

For many tasks or work packages, simple bar charts, production rate charts with trend analysis, or checklists are all that is needed, *provided* that

1.0 *General Project Information*

    1.1   The project summary plan: scope, objectives, approach
    1.2   Project appropriation requests (PAR)
            (product development, capital facilities, data processing systems projects)
    1.3   Research and development cases (R and D or product development projects)
    1.4   Product plan (product development projects)
    1.5   Contract documents (sales projects)

            1.5.1   Request for proposal and all modifications thereto
            1.5.2   Proposals
            1.5.3   Original signed contracts and modifications, and all documents and
                     specifications incorporated in the contract by reference
            1.5.4   Contract correspondence
            1.5.5   Acceptance documentation

    1.6   Statement of work
            1.6.1   X Company
            1.6.2   Y Company

2.0 *Management and Organization*

    2.1   Key organization chart
    2.2   Linear responsibility chart
    2.3   Key project personnel

            2.3.1   X Company
            2.3.2   Y Company

    2.4   Project manager and key team member job specifications
    2.5   Key functional managers and staff assigned to project
    2.6   Policies and directives

3.0 *Technical*

    3.1   Technical approach
    3.2   System specifications
    3.3   Component specifications
    3.4   Production specifications
    3.5   Drawings
    3.6   Reports
    3.7   Design review minutes
    3.8   Production plans
    3.9   Configuration status
    3.10 Engineering change notices
    3.11 Change control board minutes
    3.12 Quality assurance
    3.13 Reliability, maintainability, supportability assurance
    3.14 Field service and engineering
    3.15 Value engineering

Figure 10.21   Outline of the project file.

4.0 *Financial*

    4.1   Estimates
    4.2   Budgets
    4.3   Cost accounting reports
    4.4   Project Profit and Loss statement
    4.5   Contract status reports
    4.6   Project chart of accounts
    4.7   Billings and payment vouchers

5.0 *Works plans and schedules*

    5.1   Project breakdown structure
    5.2   Master schedule and milestone charts
    5.3   Network plans or bar charts
    5.4   Detailed schedules

6.0 *Work Authorization*

    6.1   Work orders—internal
    6.2   Work orders—other affiliated companies
    6.3   Major purchase orders
    6.4   Subcontracts

7.0 *Evaluation and Reporting*

    7.1   Project evaluation reports and charts
    7.2   Project evaluation meeting minutes
    7.3   Management reports
    7.4   Customer reports
    7.5   Trip reports
    7.6   Audit reports

8.0 *Communications*

    8.1   Internal communications
    8.2   External communications

9.0 *Project Security*

    9.1   Work classification
    9.2   Visitations
    9.3   Clearance lists

Figure 10.21   (*cont'd*)

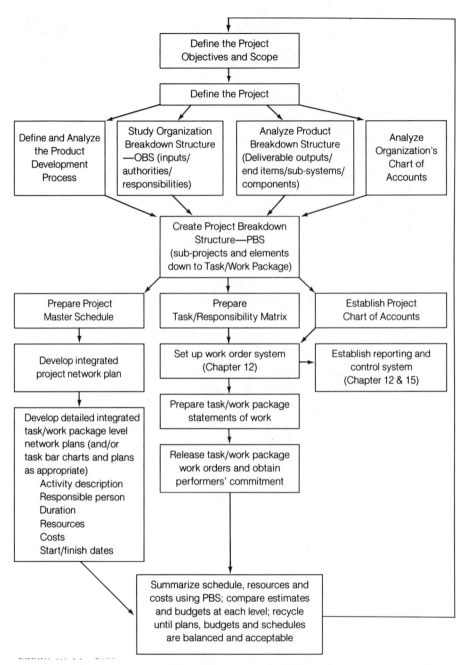

Figure 10.22 Summary of project planning steps. (Robert Youker, "A New Look At Work Breakdown Structure (WBS) (Project Breakdown Structure—PBS)," *Proceedings of the Project Management Institute Seminar/Symposium,* Calgary, Alberta, Canada, October 1990, Project Management Institute, Drexel Hill, PA, adapted from Fig. 3, p. 712.)

the key task interface events (start, complete, other) are included in the project network plan discussed earlier.

10.17   THE PROJECT FILE

The project file is an orderly collection of documents reflecting all aspects of the project. Its purpose is to assure that all pertinent information is continually available to the project manager and others regarding any matter related to the project, and it is important for a number of reasons.

- When a change in the project manager (or a key project staff member) is necessary, the project file is valuable aid in achieving a rapid, smooth transition of assignment.
- When litigation occurs or is threatened, the file provides vital information which may not otherwise have been recorded or retained.
- When a similar project is to be undertaken at a later date, it provides data for use in proposal preparation, pricing, and management planning.
- Upon project completion, the post completion audit or other study of the project file will reveal weaknesses and strengths in the unit's project management or other capabilities and thus indicate where improvements are needed.

The basic content of the project file is outlined in Figure 10.21. As the final step in the initial project planning effort, the project file is established with procedures to assure that all documents are properly retained within the file.

10.18   SUMMARY OF PROJECT PLANNING STEPS

To summarize the key points of this chapter and link it to those that follow, the steps required in planning any phase of a project are shown in Figure 10.22. At the start of the conceptual phase of a project, the depth and rigor of the planning usually will be only a fraction of that required for the definition phase, and that in turn will be a fraction of the planning effort required for the implementation or final execution phase. The same basic steps will be required during any phase, however. Each successive pass through the planning steps, as resource commitments of ever increasing magnitude are made to proceed into the next phase of the project, will result in revision and expansion of detail in the previously made plans.

11

‹ ›

Project Team Planning and Project Start-Up

As discussed in the previous chapter, many tools and methods have been developed to carry out the important jobs of project planning, including scheduling, estimating, budgeting, and the other aspects of this function identified previously. Too frequently these planning tasks have been assumed to be the responsibility only of the project manager, or even delegated entirely to specialized project schedulers and cost estimators.

In recent years, the importance of the multidisciplinary project team has been recognized more widely, and the power of project team planning has been discovered by many practitioners. This is becoming evident in the increased emphasis on systematic project start-up, using team planning workshops at the start of each phase of the project life cycle.

This chapter describes the need for project team planning, and describes methods that have been developed to carry out such planning in an effective manner. It further describes the planning deliverables which the project team can produce and identifies the important benefits that can be obtained by using the project team planning approach. The use of intensive project start-up workshops is described, using team planning methods, and a case study of their use on AT&T telecommunications projects is given to illustrate the approach and its benefits.

11.1 THE NEED FOR PROJECT TEAM PLANNING

Recognition of the need for project team planning has grown out of the increased awareness of (1) the weaknesses in the more traditional project planning approaches, (2) the difficulties in getting functional managers and team members to be committed to a plan that has been created by others, and (3) the need to accelerate the project planning and team building processes at the very beginning of a project, or at the beginning of a new phase of a project. Team planning can also be used effectively when any major change in scope is required, or when a major, unforeseen problem is encountered.

Traditional Project Planning Approaches

Traditionally, project planning activities are considered to be a primary function of the project manager. On smaller projects and within organizations that have relatively little experience in formalized project management, the project manager, if one is appointed, is typically expected to put together whatever plans and schedules may exist for the project.

In larger organizations, especially when they have considerable experience in project management and have formalized their approach to managing their projects, project planning specialists create the project plans, schedules, and budgets. The basic tools used by such specialists are described in Chapter 10. Ideally, these project planning specialists (planners, schedulers, cost estimators, cost engineers, software and computer operations specialists) will carry out their work for a specific project under the direction of the project manager assigned to that project. In other situations, they may work independently of the project manager, as discussed next.

Weaknesses in the Traditional Approaches

Several critical weaknesses often can be observed in the traditional approaches:

- **Project plans, schedules, and budgets do not reflect the realities of how the work will actually be done.** There are always many ways to plan and execute a project; even the best plans will not be followed if they do not reflect the methods that the people doing the work will actually use.
- **The functional managers and other team members, and even at times the project manager, are not committed to the plans and schedules.** If the plans do not reflect how the work will be done, the

people doing the work will not have a sense of commitment to the plans, with the result that they are not committed to the project itself.

- **More than one plan exists.** Given the above points, it is not surprising to find many situations where more than one plan exists, either for the entire project, or for many portions of it. The project manager who does not believe in and is not committed to the plans and schedules produced by a central planning department will produce his or her own plans. Many functional managers and project leaders often do the same thing.

- **The planning process is inefficient in the use of key persons' time.** A project manager or the planning specialists who recognize the need for involvement of the key project team members in creating the plans will often meet with each individual team member one-on-one to obtain the needed information. After meeting with other team members, a second or even several more rounds are usually needed to work out various conflicts or discrepancies. This process is inefficient and consumes much time of all concerned, compounding the dislike most people have for planning in the first place. This round robin approach slows down the critical start-up period of the project, and does not enhance teamwork or communication.

- **Plans created either without involvement of the key people, or with their involvement through the round robin or honey-bee approach described above, will generally be based on a bottom-up view of the project.** This bottom-up approach results in project plans and schedules which are often poorly integrated, and harbor unrecognized conflicts that will be identified later when there is insufficient time to avoid them through more integrated, top-down planning.

These weaknesses—unrealistic plans, lack of commitment to the plans, the existence of multiple plans, inefficient use of key persons' time, delay in starting up the project and building the team, and bottom-up rather than top-down planning—have led to a growing realization that there is a better way: project team planning.

11.2   THE PROJECT TEAM PLANNING PROCESS

Project Start-Up Workshops

Although the importance of getting a project off to a good, well-planned start has long been recognized (see, for example, the Project Start-Up Checklists in Appendix A and the *Handbook of Project Start-Up*,[1] prepared by the INTERNET International Project Management Association's Committee on Project Start-Up), it has only been within the last 10 years that the concept of systematic, intensive, well-planned project start-up workshops has been widely accepted and used. The INTERNET Committee has been instrumental in promulgating and documenting this concept, and the Committee's Handbook provides detailed information on the concept, the methods, and many examples of experience in its application in various industries and geographic areas of the world. Pincus[2] presents a persuasive case for using the intensive workshop approach to developing execution plans on a team basis for design and construct projects.

The fundamental essence of these systematic project start-up workshops is *project team planning*. The start-up workshop, when properly conducted, provides the setting and well-planned process that enables the project team to work together effectively to produce integrated plans and schedules in a very short time.

The phrase "project start-up" may be misleading, since the concept applies not only to the very first start-up at the beginning of the conceptual phase of a project, but also at the beginning of each subsequent phase: definition, planning or proposal; execution or implementation; and project close-out. The term "project phase transition workshop" may be more appropriate than "project start-up workshop."

Elements of the Team Planning Process

The basic elements of an effective team planning process are:

- Adequate preparation.
- Identification of the key project team members.
- Interactive exchange of information.
- Physical setting conducive to the process.
- Capture of the "Team Memory."
- Use of a planning process facilitator.

Each of these is discussed in the following sections.

Adequate preparation.   Prior to bringing together a project team for a team planning session it is vital to prepare adequately for the meeting. This includes:

- Defining the specific objectives of the team session and the results to be achieved.
- Establishing a well-planned agenda.

- Preparing sufficient project planning information in preliminary form (project objectives, scope definition, top levels of the PBS/WBS, team member list, established target schedules, if any, and so on).

- Setting the session date sufficiently in advance to ensure that all team members can attend.

- Announcing the session through appropriate authoritative channels to ensure higher management interest and support—and to assure that all team members show up.

- Defining and understanding the planning process to be used, and the roles and responsibilities of the project manager and the planning process facilitator.

- Arranging for a suitable meeting facility and related logistical support.

Pincus states that "The workshop approach should not be considered without the full endorsement and participation of the project manager. The project manager must control the planning and decision-making."[3]

Identification of the key project team members.   In order to have a project team planning session, it is necessary to identify the team members. However, this is often not a simple task. Who are the *key* project team members? Which functions must be included, and what level of manager or specialist from each function should be identified and invited to the team planning session? How many people can participate effectively in such a session?

A few basic rules can be helpful in answering these questions:

- Each of the important functional specialties contributing to the project must be represented. This may include people from within and outside of the organization (contractors, consultants, major vendors, customer representatives where appropriate, and so on.)

- The persons holding responsibility and accountability for the project within each functional area (the functional project leader) must be present.

- If a functional project leader cannot make commitments of resources for his or her function, that person's manager (who *can* make such commitments) should also be invited to participate in the team planning session.

- If the key team members number more than 20 people, special efforts are needed to assure appropriate interaction (such as plenary sessions combined with breaking into smaller working team sessions).

- The project manager obviously plays a vital role in the team planning sessions; and, if available and assigned, project planning, scheduling, and estimating specialists should also participate in—but not dominate—the sessions.

Interactive exchange of information.   Central to the project team planning concept is the need for intensive interaction between the team members during the planning process. The session preparation, the information provided, the physical setting, and the methods of conducting the planning sessions must be designed to promote, not inhibit, this interaction. If the project manager goes too far in preparing planning information prior to the meeting, and presents this information as a *fait accompli* to the team in a one-way presentation, there will be little or no interaction, and the objectives of the team planning session will not be met.

Gillis has pioneered the development of such sessions over several decades in his work in Canada, the United States, and Europe, and Figure 11.1, from one of his papers,[4] illustrates several important factors in achieving the interactive exchange of information that is needed for effective project team planning. As indicated in Figure 11.1, the interaction process is based on:

- Immediate recording of keyword abstracts of what is said (recall trigger).
- Immediate display of the group memory.
- Exploration of what it means (through interactive discussion).
- Fitting the keyword card in the right information structure.
- Continuing the process until the objectives of the session have been reached.

In developing the project master schedule by using the PBS and the major milestones and interface events which have been identified, very effective integration of all functional tasks can be achieved. Serious conflicts are quickly revealed, and become the subject of later, more detailed planning and analysis involving only the specific parties concerned.

Physical setting conducive to the process.   As shown in Figure 11.1, Gillis uses the term "planning theater" for the room in which the interactive team planning sessions are held. Such a facility is an important factor in achieving the interaction desired. It does not have to be an elaborate design, but it must provide:

- Plenty of wall space with good lighting for display of the team memory and planning results.
- Open access to the walls by the team members (elimination of large tables and other impediments to individual movement and interaction to fill in keyword cards and place them on the walls).

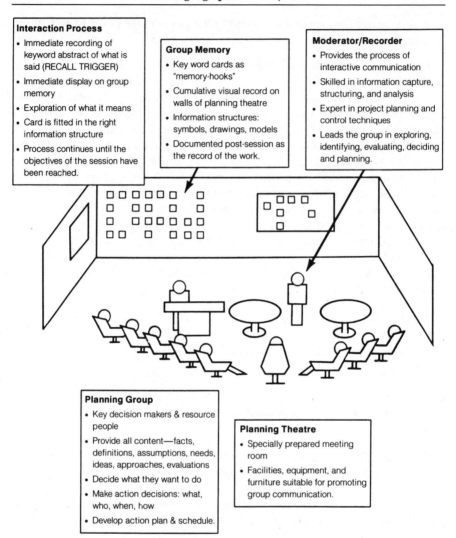

Interaction Process
- Immediate recording of keyword abstract of what is said (RECALL TRIGGER)
- Immediate display on group memory
- Exploration of what it means
- Card is fitted in the right information structure
- Process continues until the objectives of the session have been reached.

Group Memory
- Key word cards as "memory-hooks"
- Cumulative visual record on walls of planning theatre
- Information structures: symbols, drawings, models
- Documented post-session as the record of the work.

Moderator/Recorder
- Provides the process of interactive communication
- Skilled in information capture, structuring, and analysis
- Expert in project planning and control techniques
- Leads the group in exploring, identifying, evaluating, deciding and planning.

Planning Group
- Key decision makers & resource people
- Provide all content—facts, definitions, assumptions, needs, ideas, approaches, evaluations
- Decide what they want to do
- Make action decisions: what, who, when, how
- Develop action plan & schedule.

Planning Theatre
- Specially prepared meeting room
- Facilities, equipment, and furniture suitable for promoting group communication.

Figure 11.1   Team planning session using a planning theatre. (Source: R. Gillis, "Strategies For Successful Project Implementation," INTERNET *Handbook of Project Start-Up, op. cit.,* Section 4. Used by permission.)

- Sufficient space to enhance individual comfort, movement to view other walls, and open communication.

Capturing the "team memory."   The team or group memory is based on:

- Using the keyword cards as "memory hooks" to recall specific ideas.
- Creating a visual record on the walls of the planning theater or meeting room.
- Proper information structures that are appropriate to the planning work being done: models, matrices, drawings, symbols, charts.

The team memory resulting from capturing and structuring the information exchanged and produced during the planning sessions provides an agreed, understood post-session record of the plans created by the team.

More teams are beginning to use personal desktop or laptop computers during the planning sessions, with screen projection to enable the team interaction. This enables capture of the team memory in a very usable form, and each team member can leave the workshop with hard copies of the results, produced on site. Whiteboards that can produce xerographic copies are also very productive tools for generating project breakdown structures and master schedules. 3M "Post-Its,"® the adhesive pieces of paper that are available in various colors, are also very useful during project team planning sessions. Some organizations have printed up special Post-It forms designed for their specific planning purposes.

Use of a planning process facilitator.   The facilitator (moderator/recorder in Figure 11.1) is a crucial player in the project team planning process. This person:

- Provides the process of interactive communication.
- Is skilled in information capture, structuring, and analysis.
- Is expert in the application of project planning and control methods and techniques and other aspects of project management.
- Leads the team during the planning sessions in the processes of exploring, identifying, evaluating, deciding, and planning.
- Maintains the process discipline to adhere to the established agenda for the planning session.

The Project Manager Role in Team Planning

It is widely recognized that the key characteristic of the project manager role is that of *integration*. In project team planning, the integrative role of the project manager becomes quite obvious. The project manager must assist the project team members in developing acceptable plans and schedules that achieve the objectives of the project and reflect the plans and available resources of the various team members. The project manager usually has the lead responsibility for preparing for the team planning sessions.

During the planning sessions, the project manager must be alert to real or potential conflicts in plans, and bring these to the surface for resolution. He or she must concentrate on identification of the key project interface events, or those points of transition of responsibility from one team member to another, since one of the key tasks of the project manager is to properly manage these interfaces.

A second key role in the project team planning process, that of the process facilitator, can also be taken by the project manager, but experience shows that it is much more effective for another person to carry out the facilitator role.

Setting the Stage for Detailed Planning

The plans, schedules, and other planning documents created during a team planning session should be limited to integrated plans at the overall project level. These will require definition of the project down to the major functional task level, so that responsibilities can be assigned, agreed, and understood among the team members. The team-produced project master schedule will show the agreed target dates for key milestones, reflecting the team's judgment on the overall allocation of time to accomplish the intermediate and final objectives.

Team planning sessions are not intended to produce detailed, functional task plans, schedules, and budgets. To attempt to do so would be an extravagant waste of time. Rather, these team sessions are intended to set the stage for truly effective detailed planning, scheduling, and budgeting. Based on the results of the top-down planning performed by the project team, under the integrating influence of the project manager and guided by the planning process facilitator, the stage is set very effectively for the detailed planning needed to validate the team's efforts and prove whether their judgments are correct.

At this point, the project manager, often with the assistance of planning, scheduling, and estimating specialists, can proceed with the more detailed, integrated planning that is necessary to assure effective monitoring and control of the project. The top-level project plans produced by the team can be entered into the computer software to be used on the project, often during the team planning sessions themselves. The more detailed functional task plans and schedules can also be entered in the planning and control system, usually following the team planning sessions, to the extent that this is warranted and practical.

11.3 PROJECT START-UP WORKSHOPS IN THE TELECOMMUNICATIONS INDUSTRY: A CASE STUDY[5]

This case study summarizes the experience within one part of AT&T in the use of project team planning and project start-up workshops. Within a company the size and scope of AT&T, with annual sales of over $40 billion in diverse products and services, there is a wide variety of projects of all sizes, and of degrees of risk, complexity, and character of the end result. This case study deals with projects in which the Business Communication Systems (BCS) Group of AT&T, with sales of over $2 billion per year, has agreed under fixed price contract to companies, institutions, and agencies that it will design, manufacture hardware, develop software, install, test, and cutover into operation complex, high-technology voice/data telecommunications and related systems. Such projects usually must be completed within a few months to perhaps a year, although some multiproject contracts cover several years.

Within AT&T BCS across the United States, there will be many projects of this type under way at one time, ranging from small switchboards to systems covering an entire state government. There will also be a small number of mega-projects for federal, state, and local government agencies or nationwide corporate facilities. Such projects must be executed so that the new facilities are in place and tested to enable a rapid cutover from the old to the new, usually over a weekend, within minimum disruption to the on-going operations of AT&T's customer. Achieving the agreed cutover date is almost always of extreme importance to the customer, and once the cutover date has been agreed, it is very difficult to change. Cost is fixed by the contract, but changes in project scope often affect the total system cost. The performance specifications to be met by the system are spelled out in great detail in the contract terms.

Project Management in AT&T Business Communications Systems

Over a number of years, AT&T has been moving toward more formalized project management practices, and in this business sector there is a National Director of Project Management holding responsibility for the management of all BCS contractual projects. The National Project Director has three project directors (each responsible for a region of the country) and specialized staffs reporting to him. The three project directors have a number of program and project managers and project scheduling specialists reporting to them. A full-time project manager is assigned to a project when its value exceeds a few million dollars. Some projects exceed a value of $100 million. In some cases, a project manager is assigned for smaller contracts if the project is unusually complex, either technically or organizationally. Other exceptions are made for smaller projects which are parts of a larger program. A program manager holds responsibility for multiproject contracts with one customer, and usually has several project managers or site managers (a step below project manager) reporting to him or her.

Because of the rapid growth of this business segment within AT&T, a cadre of experienced project managers was not available, and there has been a continual need to indoctrinate and train new people, who are experienced within other parts of AT&T, in the requirements of the project manager assignment and project planning and control principles, practices, and tools. As a result, AT&T has established a broad ranging education and training program in project management, and has stressed the certification of its project managers and supporting staff as Project Management Professionals with the Project Management Institute (PMI).

The AT&T project manager operates within a classic matrix organization, usually as a one-person project office, but using specialized staff support as required. In very large projects, the project manager will have several people on his or her direct staff. Many different parts of AT&T located in different geographic regions must contribute to each of these projects: several engineering and technical disciplines, purchasing, manufacturing, field installation and test, provisioning and logistic support, software development, training, and various other services and operations departments.

Identification of the Need for Project Start-Up Improvement

In the mid-1980s, the then Director of Projects for Southern California and Hawaii identified a need for ways to accelerate the planning, learning and teambuilding processes which take place on every project. He saw this need for his own project managers as well as the functional managers who carried out the specific tasks on each project. Very importantly, he also saw this need within the customer's people who were involved with the project. In addition to this acceleration, he wanted to prevent or quickly resolve any adversarial attitudes that may be encountered at the start of a project with the customer's assigned people. Such attitudes, when encountered, take some time to overcome and had sometimes caused delays and added costs.

Typically, after a new project had been under way for a few months, good teamwork emerged. The Director of Projects wanted to achieve that teamwork in a few days or weeks, due to the short duration of many of his projects.

Satisfying the Need with Project Start-Up Workshops

The Projects Director decided to initiate project start-up workshops on his new projects to see whether these needs could be met in this manner.

Approach.   A three-day start-up workshop format was designed. The first two days, spaced at least a week apart, involved only AT&T people. The third day, following the second day by at least a week, included the customer people who were involved directly in the project and also senior customer managers.

Workshop objectives.   The start-up workshop objectives were:

1. To apply proven project management methods to the project, and develop—as a team—jointly agreed project plans, schedules, and control procedures.
2. To assure good understanding of the roles and responsibilities of all AT&T and customer project team members, thereby enhancing effective teamwork.
3. To identify additional steps needed to assure project success.

While team building was not stated specifically as an objective, it obviously was one of the most important results to be achieved.

Start-up workshop planning deliverables.   Emphasis throughout the workshop sessions was on the deliverables to be produced by the team. These were:

1. Agreed list of key project team members.
2. List of key concerns and major open issues.
3. A well-defined project/work breakdown structure (PBS).
4. A task/responsibility matrix, based on the PBS, and reflecting all identified contributors to the project, including the customer and outside agencies (such as the involved local telephone operating companies).
5. A list of key project interface events, linked to the PBS and showing the initiator and receiver(s).

6. A project master schedule, based on the PBS, reflecting the key project interface events, and based on the consensus of the project team on the overall allocation of time.

7. Agreed procedures for project monitoring and control, including dates for periodic project review meetings.

8. Action items resulting from the start-up workshop discussions, with assigned responsibility and agreed due date for each.

Workshop agenda and conduct.   The agenda for the first two days covered these topics:

First Day

1. Introductions and Workshop Objectives
2. Project Overview, Scope and Open Issues
3. Description of the AT&T Project Management System to Be Used
4. Systematic Project Definition
   —Concept
   —Team Assignment: PBS
5. Task/Responsibility Matrix
   —Concept
   —Team Assignment

Second Day

1. Project Interface Management
   —Concept
   —Team Assignment: Key Interface Event List
2. Project Scheduling and Time Control
   —Concept
   —Team Assignment: Project Master Schedule
3. Project Monitoring and Control
4. Action Assignments.

At least one week between the first and second days was required to enable the project manager to clean up the results of the first day for use in the second session. One week was similarly provided before the session with the customer team members, for the same reason. The agenda for the customer session covered the following topics:

Third Day (with Customer)

1. Introductions and Workshop Objectives
2. Project Overview, Objectives, Scope, and Customer Objectives

3. Description of the AT&T Project Management System to Be Used
4. Presentation and Modification (as Required) of the Implementation Plan
   —Project Team List
   —Project Breakdown Structure
   —Task/Responsibility Matrix
   —Key Project Interface Event List
   —Project Master Schedule
   —Monitoring and Control Procedures
5. Action Assignments.

Depending on the situation, these topics may be covered in two hours, four hours, or a full day. The primary objectives of this session were (1) to get the customer's inputs, involvement, modification of, and commitment to the plans; (2) to show the customer that a well-conceived, achievable plan and schedule has been established; (3) to assure that the customer knows what is expected of him or her in order to achieve the project objectives; and (4) to establish an early client perception that AT&T is the expert in these types of installations. In most of these AT&T projects, the customer has significant responsibilities which, if they are not carried out on schedule, will jeopardize project success.

Conduct of the sessions. The workshop sessions are the responsibility of the assigned project manager. He or she plans and prepares for the sessions, with the assistance of a staff or outside consultant acting as the facilitator. For each topic listed in the agenda, the consultant briefly presents the underlying concept to be applied. Then the project team members roll up their sleeves and create the deliverable item for the project: the PBS, task/responsibility matrix, key project interface event list, and project master schedule.

Some of these items, especially the PBS, matrix, and interface event list, are usually developed by breaking into 5 or 6 person teams, with each team covering assigned parts of the project. These small teams then report their results back to the full project team (usually 15 to 20 people), to assure total team buy-in of the plans.

In this process, the consultant acts as a facilitator, assures that the overall process is adhered to, and is a source of industry-proven project management knowledge.

One of the overall objectives of the start-up workshop is to position the project manager properly in the eyes of the other AT&T team members, and also of the customer team members. The project manager thus must be seen as basically running the start-up workshop sessions, with the assistance of the facilitator.

The Project Master Schedule

This important planning deliverable is developed with the entire team using a timed-scaled wall chart display, usually a four- by six-foot chart. The level 2 and 3 PBS elements are written on the left side, and key milestone and interface events are identified and placed at the appropriate points on the calendar scale. Level 4 and 5 elements and tasks are usually shown as bars on the master schedule. Small 3M Post-Its® are useful in this process to depict the milestone and interface events. Some project managers have used their personal computers with large screen projection to develop the master schedule with the team.

The project master schedule becomes the top level of the schedule hierarchy for the project. After the start-up sessions, the project manager develops the detailed project action plan using PERT/CPM/PDM methods and software on his or her microcomputer. This validates the master schedule, shows where changes are required in it, and enables effective monitoring and control at both the detailed and overall project levels.

The AT&T Project Management System

A description of the overall management system to be used for the project is given to the AT&T team members on the first day, and to the client team members on the third day. This is a descriptive overview of the principles, methods and information systems to be used. An important integrative element of this system is a proprietary data base which brings together technical and management information on the project. Extensive use of the latest laptop computer technology is made in building and interrogating this data base system.

Results Achieved

The most direct indication of the overall benefits of using a well-organized process for starting up projects is that the system cutovers—project completions—have been on schedule and with better quality on projects using this approach compared to the projects that did not.

Better project and functional planning.   The start-up workshops get the project team started quickly, with a good understanding of *what* needs to be done, *who* does each of the many tasks, and *when* each must be completed. This approach gets all of the functional organizations thinking about what kind of planning is required—before getting into the thick of the action. Previously, some functional managers would leave the planning to the last minute, or would not do any planning at all.

Better communications and teamwork.   After the start-up workshops all project team members use the same semantics and planning terms. By jointly working through the planning deliverables, good teamwork is achieved much earlier on each project. This joint planning shows each team member that everyone on the team has important tasks to perform, and how these tasks interrelate. There is a better realization that they all need to be involved in the planning effort to assure project success.

Improved customer relations.   There have been very positive reactions from customer team members and higher management to the start-up workshop sessions and the resulting deliverables. AT&T marketing managers have given similar positive reactions, and point to the fact that the workshops provide a vehicle for the AT&T team members to work closely with the customer team members very early in the project. The adversarial attitudes that have previously been experienced on some projects have been avoided. An important result of the third-day session with the customer team members and managers has been quick escalation and resolution of open issues that threaten to delay the cutover.

Benefits to the project manager.   Several important benefits to the AT&T project manager have been noted:

- **Positioning the project manager.** Typically, the marketing people have been working on the sale to a particular customer for months if not years, preparing the proposal and negotiating the contract. The project manager, who is often involved during the proposal preparation stage, takes over implementation when the contract is signed. Before using the start-up workshops, it would usually take the project manager some time to establish his or her position with both the customer and AT&T team members, especially the marketing people who naturally feel a strong proprietary interest in the project. By the end of the start-up workshop process, all team members have a good understanding of the need for and the role of the project manager, and are ready to give him or her the required support.
- **Detailed planning and scheduling.** Another benefit to the project manager is that he or she can immediately use the project breakdown structure, the project master schedule, and the project interface event lists produced by the team as the basic framework for the detailed PERT/CPM/PDM network plan and schedules.
- **The project manager as project interface manager.** Another benefit is the understanding by the team members of the role of the project manager as the project interface manager (see Chapter 13). By getting the team to identify the key project interface events, and explaining to

the team members how the project manager manages these project interface events while each functional manager manages his or her assigned tasks, good acceptance of the project manager role has been achieved. The functional managers quickly realize that this is a valuable help to their getting the job done successfully: to specification, on schedule, within budget.

Key Roles in the Start-Up Workshop Process

The project manager and the facilitator carry the most important roles in the start-up workshop process. In addition, a third process role has been found to be very useful, and has been added recently—that of an associate project manager, who serves as the workshop recorder/sergeant-of-arms. These three roles are briefly discussed next.

The project manager's role.   The project manager holds the lead responsibility for getting the project started efficiently. Prior to the first kickoff workshop session, he or she has substantial work to prepare the materials, arrange the meeting logistics, identify and invite the team members, and insure that they will attend. The project manager exercises overall control of the kickoff sessions, reinforcing his or her position with the project team. He or she handles all project-specific references and decisions during the sessions, with the active involvement of the marketing and other functional project team members.

The start-up process facilitator's role.   The facilitator is the authoritative source of information on project management principles and practices. He or she handles all the generic descriptions of the methods to be applied to the project and their underlying principles. The facilitator takes on any resistance to the use of the methods and provides the reasons for using those methods. The intention is to present the image that anyone disputing the methods is arguing with the project management body of knowledge, and not with AT&T, or one division of AT&T, or the project manager.

In addition to being the project management resource person, the facilitator assures that the start-up process is followed effectively. This means following the agenda reasonably closely, using the team methods which have been found effective, and transferring experience from one project start-up workshop to the next.

The associate project manager's role.   The third process role, which has recently been added on larger projects, is an assistant to the project manager during the workshop sessions. This person is a project management staff member who may not be assigned to the project in question, but who is experienced in project management and the start-up process. He or she acts as a recorder during the sessions and is encouraged to use a laptop personal computer (which is commonly used for several purposes on projects within AT&T BCS) to accelerate the production of the workshop deliverables. The project manager, with this assistance, can concentrate on running the meeting, developing the information and workshop deliverables, and assuring that all team members understand what is going on.

A second important function of this process aide is that of sergeant-of-arms. In this role, the person observes the conduct of the team members and calls attention to inappropriate meeting behavior: side meetings, getting into problem-solving sessions, getting off the subject, leaving the workshop for telephone calls or other unwarranted reasons, and so on. By relieving the project manager of these duties, there is less opportunity for the project manager to engender antagonism during the formative stages of project team building.

The functional project team members' roles.   The functional team members' roles in the kickoff workshops are vital. These are the people who will be planning and executing the functional tasks necessary to complete the project. Their primary roles in the workshops are: First, to provide the specific knowledge and expertise, within each of their areas of specialization, needed to create the workshop deliverables; and second, to make commitments for their functional organizations regarding responsibilities and schedules.

The Hidden Agenda Items within the Start-Up Workshop Process

In addition to the deliverables and other results described above, there are several important "hidden agenda" topics and related results involved in the start-up workshop process. The most important of these—not listed necessarily in order of importance—are:

1. Introducing uniform, proven industry project management practices with common terminology.
2. Providing hands-on training to all project team members in effective project planning and control methods.
3. Tapping the wisdom of the group—"group-think"—to develop the best overall project plan.
4. Creating a shared vision of the total scope of the project, its challenges and its objectives at several levels.
5. Demonstrating the power and benefits of open team communications.
6. Exchanging experience and developing planning skills and understanding among the team members of all aspects of what goes into a complex AT&T telecommunications-information systems project.

7. Creating a working team, and getting individual team member commitment to, and enthusiasm for, the project—through involvement, understanding, and commitments made to the peer group, and not just to the project manager.

The approach used—short presentations, small team assignments, and team reports to the full group—has proven to be effective in exchanging and transferring knowledge and experience. Using this approach, individual and team development takes place at four levels:

1. The facilitator and project manager convey a certain level of knowledge with the initial presentation on a given topic.
2. The members of the small teams work together for periods of one or more hours, and interchange ideas and experience; there will always be diverse levels and types of experience and knowledge in each small team, and the team members learn from each other.
3. As each team reports its results to the full project team, exchange and learning takes place as the team members see what the other small teams have done. Team members also get the experience of making presentations to the full team.
4. The workshop facilitator and the project manager add to and expand on the information presented—or show where it may need further improvement. Additional knowledge transfer and personal development occurs during these discussions.

Lessons Learned

The most critical lesson learned after starting up a very large number of major projects in the manner just described is that thorough preparation by the project manager prior to the first workshop is crucial to success. The project manager, with appropriate assistance, can develop the preliminary deliverables (PBS/WBS, project master schedule, responsibility matrix, etc.) and thereby save much time in the workshops. However, if he or she goes too far with this and presents them as completed items, the team members will not become involved and will not "buy in" to the degree considered necessary.

A second lesson learned is that adequate lead time is needed to assure that all key team members will be able to attend. In some cases, it is necessary to apply management pressure to get the needed priority applied to the start-up workshops, so that schedules can be cleared of other matters.

A final important lesson is that achieving the start-up workshop objectives takes time, and if adequate time is not allocated to this process, the desired results will not be achieved. The team members should be prepared to spend a long day or days in the workshops, if required. Additional follow-on effort is required by the project manager and the team members to validate and complete the deliverables prepared in the workshop sessions.

Modifications Made for Smaller Projects

In order to achieve the same objectives on smaller projects, and on larger projects that are very similar to or use the same team members as other recent projects, the following modifications have been made to the procedures described earlier:

1. Reduce the duration of the internal sessions from two days to approximately four hours, starting at 10:00 AM so that team members can attend to on-going responsibilities before and after the workshop.
2. Condense the formal presentations to fit the time.
3. Prepare drafts of the project breakdown structure, the task/responsibility matrix, the interface event lists, and the project master schedule ahead of time, and incorporate any changes introduced by the team members.
4. Shorten the client workshop to four hours or less.

The objectives, preparation, invitations, and workshop deliverables for smaller projects are the same as for large projects. If the same project workshops are conducted with the same professionalism as large projects, the project manager can instill the same team spirit, generate the same quality plans, and use the same monitoring of progress and quality while demonstrating a sensitivity to the functional managers' workloads.

Conclusion

The start-up workshop process described here has produced beneficial results by bringing the project team members together early, and by concentrating intensively on a few basic fundamentals of effective project management. AT&T is committed to continued use and improvement of these project start-up workshops.

11.4 BENEFITS AND LIMITATIONS OF PROJECT TEAM PLANNING

The basic benefits of project team planning are:

*   The plans produced will be based on how the work will actually be accomplished.

- The persons responsible for performing the work will have a greater sense of commitment to the plans and to the project.
- Only one set of plans will exist: Those that the project team has created and is following.
- The time required to be devoted to planning by the key project team members will be minimized.
- The project plans will reflect a top-down approach using the total wisdom of the project team, which then sets the stage for more effective, detailed, bottom-up validation of the plans.

Limitations of Project Team Planning

The decision to use project team planning should be based on the characteristics of the project in question. If it is an effort that is very well known to the organization, and very repetitive of many previous projects, with project team members who are all experienced in this type of project, and planners who can produce plans and schedules that are valid and acceptable to all concerned, then it would not be appropriate to insist on the type of project team planning described above.

There appears to be no upper limit in project size for the use of the project team planning approach. At the top of a massive mega-project, as one extreme, the objective of the top level project team planning session would be to define the major subprojects into which the mega-project should appropriately be divided, identify the key milestones and interface events which will link these subprojects, assign responsibilities as appropriate, and lay out the target project master schedule. At each subordinate level, the project team must recognize the appropriate level of detail below which they must not attempt to develop plans and schedules. Each team must concentrate on handing down the structured plans and schedules within which the next level teams must in turn develop their plans.

The primary limitation in project team planning is probably the time required of the project team members to devote to the team planning sessions. Although planning should be given a high priority in any organization, frequently planning is viewed as unproductive and even wasteful, hence it is difficult to convince the project team members that they should devote even a few days to developing the project plans. Top management understanding and support is required to overcome these ingrained attitudes and habits. A successful project team planning session can also do a lot to demonstrate the power and usefulness of this approach.

12

‹ ›

Controlling the Work, Schedule, and Costs

I f the planning actions described in Chapters 10 and 11 are properly carried out, and the project and task plans, schedules and budgets are well documented, then it will be possible to exert good control over the work, schedule, and costs. Simultaneously, technical progress monitoring and performance measurement are required to enable overall evaluation of the project.

The definition of project control given in the introduction to Chapter 10 should be reemphasized here. The project and functional project leaders achieve cooperative control by:

- Establishing a joint understanding of the project objectives and goals.
- Jointly defining, planning, scheduling, and budgeting the tasks.
- Using established procedures to authorize the work, to control changes and scope of work, and to control schedules and costs (as described in this chapter).
- Measuring and evaluating performance in cost, schedule, and technical terms on a joint basis to identify current or future variances from plan and to initiate appropriate corrective actions (as described in this and the following chapters).

12.1   WORK AUTHORIZATION AND CONTROL

After a project has been defined, planned, scheduled, and budgeted, effective project management requires that the individual tasks be communicated in *written, documentary form* to the persons who will direct and perform the work, authorizing them to expend money, manpower, and other resources on the project. This authorization will specify the required schedule and agreed budget. The project manager must obtain agreement and commitment (preferably by signature) that each manager or supervisor accepts the assignment. Such agreements may of course be modified, revised, or cancelled when conditions change.

This process is referred to as work authorization and control. It is required for tasks performed within the sponsoring organization and also by outside contractors, suppliers, or vendors. The general flow of work authorization is:

- Contract is awarded (or PAR, research and development case, or other go-ahead approval granted), and master contract release is issued.
- Project release document is issued.
- Project and task schedules and budgets are revised as necessary and time phased to reflect the date of go-ahead and changes made during the approval process.
- Task work orders are agreed upon and issued for all initial tasks.
- Subcontracts and purchase orders are issued as required.
- Cost accounts are authorized and opened for active tasks; budgets are entered into EDP system (if used).

At this point, work is initiated. New work orders, subcontracts, and purchase orders are issued as their scheduled start dates approach. Expenditures are recorded against the proper cost accounts for monitoring and control purposes. When a task or subtask is complete, its cost account is closed and further charges to it are rejected, unless special late charge authorization is made.

Figure 12.1 illustrates the work authorization flow described and shows the correlation of the functional organization with the Project Breakdown Structure.

The Contract and Project Releases

Upon contract award, the *master contract release* (or equivalent) document is prepared by the contract administrator and distributed to the key managers concerned.

In one company, as an illustration, this document consists of 12 pages, with appropriate attachments:

1. Summary sheet.
2. Statement of work.
3. Items and prices—hardware.
4. Items and prices—data requirements.
5. Drawings.
6. Delivery schedule.
7. Inspection, packaging, shipping, and billing data.
8. Replacement orders.
9. Property requirements.
10. Tooling and test equipment.
11. Contract requirements.
12. Support requirements.

Additional or alternative items would be required for other companies.

The *project release* document is issued next. This authorizes the total funding of direct costs for the actual execution of the project. Although it may carry different names in various organizations, its purpose is still the same. For new product development projects, for example, the research and development case document itself may serve the purpose of the project release, with appropriate approvals or cover sheet.

Figure 12.2 illustrates a typical project release document. This requires approval of the project manager, contract administrator, director of operations or manager of projects, and comptroller.

If additional funds are required to complete the project because of unforeseen problems or a change of scope, a revised project release is required prior to expenditure of such funds. In overrun situations a *management release* is used to provide the needed funds either by reducing the gross margin on the contract or from some other source.

Task Work Orders

A formal procedure is required for authorizing and controlling tasks and their related budgets and schedules. The key document in this procedure is usually referred to as a *work order*. Frequently, companies have a rigid work order procedure for relatively trivial expenditures, such as building maintenance, but no equivalent procedure for authorization and control of much larger project funds, including research and development cases.

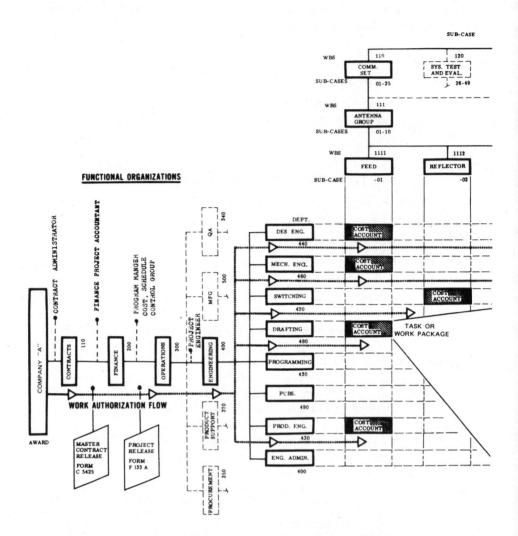

Figure 12.1 Integration of functional organizations and project/work breakdown structure. (Source: ITT Defense Communications Division. Used by permission. See also Fig. 10.18.)

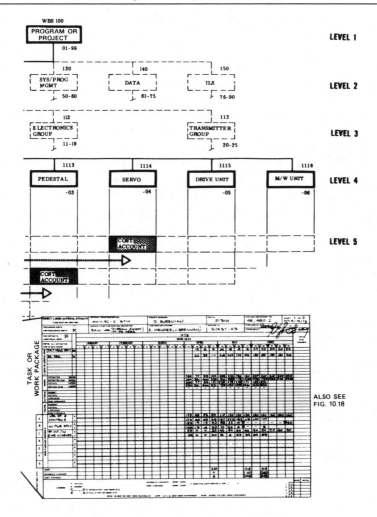

ALSO SEE
FIG. 10.18

PROJECT RELEASE

ISSUE NO. _____ ACCOUNT NO. _____

CONTRACT NO. _____ TYPE _____ DATE OF ISSUE _____
CONTRACT DATE _____ PRODUCT CLASS _____ DEPT. NO. _____
CONTRACT COMPLETION DATE _____ PROPOSAL NUMBER _____
PROGRAM/PROJECT MANAGER _____ CONTRACT ADMINISTRATOR _____

TITLE/QUANTITY _____

TOTAL AUTHORIZED COSTS

DIRECT LABOR
   1. ENGINEERS .... $ _____
   2. TECHNICIANS .... _____
   3. DRAFTSMEN AND TECHNICAL WRITERS .... _____
   4. MACHINISTS .... _____
   5. BENCH MACHINISTS .... _____
   6. CLERICAL .... _____
   7. TEST AND INSPECTION .... _____
   8. OTHER .... _____
      TOTAL DIRECT LABOR .... $ _____

DIRECT LABOR - FABRICATION
   4. MACHINISTS .... $ _____
   5. BENCH MACHINISTS .... _____
   6. CLERICAL .... _____
   7. INSPECTION .... _____
   8. OTHER .... _____
      TOTAL DIRECT LABOR - FABRICATION .... $ _____

OVERHEAD - OTHER (%) .... _____
OVERHEAD - FABRICATION (%) .... _____
DIRECT MATERIAL .... _____
SUBCONTRACTED SERVICES - MATERIAL .... _____
SUBCONTRACTED SERVICES - OTHER .... _____
PREMIUM TIME .... _____
       COMPANY CHARGES - MATERIAL .... _____
       COMPANY CHARGES - OTHER .... _____
TRAVEL .... _____
SPECIAL FACILITIES - TOOLS AND TOOLING .... _____
SPECIAL FACILITIES - CAPITAL .... _____
OTHER .... _____

      TOTAL AUTHORIZED FOR EXPENDITURE .... $ _____

      LESS: MANAGEMENT FUNDING PER RELEASE NO. .... $ _____

      TOTAL CUSTOMER FUNDED COSTS .... $ _____

REMARKS _____
_____
_____
_____

APPROVAL _____ APPROVAL _____
      PROGRAM/PROJECT MANAGER   CONTRACT ADMINISTRATOR

_____   _____
      DIRECTOR OF OPERATIONS   COMPTROLLER

Figure 12.2   Example of a project release document.

This is an obvious weakness in their ability to control a project's scope, schedule and budget. Work orders should contain at least the following:

- Brief, but complete, *statement of work.*
- Relationship to the *project breakdown structure,* with appropriate code numbers (if used) for the work package and parent element in the project breakdown structure.
- *Budget,* divided into labor and material dollars, labor hours and quantities, and other direct costs by time period, if appropriate.
- *Schedule,* including task start and completion, and known intermediate milestone and interface events, with indication of interfaced (interrelated) work control packages.
- Reference to applicable product *specifications,* drawings, and other documents.
- *Cost account code,* with provisions for subaccount codes for control of further delegation of assist work, if possible and desirable.
- *Signature of the initiator* (the project manager or a task manager for subsequent assist work orders).
- *Signature of person authorized to accept responsibility* for performance of the work as specified and his organization identification name or code (e.g., functional project leader or task leader).
- Any *special terms or conditions.*

Figure 12.3 is an example of a typical work order document. As indicated, the task schedule and budget, plus other appropriate documents, should be attached.

Lower level authorization of work is frequently needed. Subordinate work assist orders, releases, fabrication orders, shop orders, purchase orders, requisitions, test requests, and so on, should be issued or approved by the person holding responsibility for the basic work order. Provision for cost accumulation back to the basic work order cost account number is required.

Subcontracts and Purchase Orders

Formal authorization to perform work and/or expend funds externally is usually required by established policies. Procedures and documents are well established in this area. The project manager must understand and know how to use the company's subcontracting and purchasing departments and procedures to assure effective project control.

ENTER X X WHERE ENTRY OR SIGNATURE NOT REQUIRED

	GIVER	CONTRACT NO.	PROGRAM/PROJECT NAME				CONTROL LEVEL	SUMMARY ACCOUNT NO.
TASK WORK ORDER			TASK TITLE (27 SPACES)					DELIVER TO
	RECEIVER	REF. NO.	CONT. ITEM NO.	TASK NO.	START DATE	COMPLETION DATE		

ALL CLASSIFIED MATERIAL REQUIRED TO SUPPORT THIS FORM MUST BE HANDLED IN ACCORDANCE WITH PROCEDURES. THIS FORM IS NOT TO BE CLASSIFIED, NOR ATTACHED TO ANY CLASSIFIED MATERIAL.

SECURITY

	INSPECTION POINT	CONTRACT–PECULIAR INSPECTION REQUIREMENTS:
QUALITY ASSURANCE		

STATEMENT OF WORK: (IF THIS SPACE IS INADEQUATE, USE ATTACHMENT(s))

SUPPLEMENTS

1. STATEMENT OF WORK
2. SCHEDULES
3. BUDGET
4. DATA REQUIREMENTS
5. GIVER–RECEIVER PLAN
6. SPECIFICATION REQUIREMENTS
   A. APPLICABLE DOCUMENTS
   B. REQMTS – PERFORMANCE
   C. REQMTS – DEFINITION
   D. REQMTS – DESIGN–CONSTR.
7. PRODUCT EFFECTIVENESS
8. ACCEPTANCE REQUIREMENTS
9. SCN FORM
10. PROVISIONABLE ITEMS LIST
11. OTHER

CODE APPLICABLE

U – UNCLASSIFIED ATTACHED
C – CLASSIFIED SEPARATE COVER
N – NOT APPLICABLE

SIGNATURES AS REQUIRED – * INDICATES MANDATORY

REVISION

BUDGET

$(000)

ORIG

A

B

C

D

E

GIVER

INITIATOR	PROD. EFF.*	APPROVAL *	DATE	CONTROL * POINT	DATE
					OUT
					IN
					OUT
					IN
					OUT
					IN
					OUT
					IN
					OUT
					IN
					OUT
					IN

RECEIVER

CONTROL * POINT	DATE	APPROVAL *	DATE
	IN		
	OUT		
	IN		
	OUT		
	IN		
	OUT		
	IN		
	OUT		
	IN		
	OUT		
	IN		
	OUT		

SUMMARY ACCOUNT NO.

DISTRIBUTION: WHITE RECEIVER; CANARY REC. CONTROL POINT; PINK GIVER CONTROL POINT; BLUE GIVER; GREEN PROD. EFF.

Figure 12.3 Example of a work order document.

12.2 THE BASELINE PLAN, SCHEDULE, AND BUDGET

The project baseline plan, schedule, and budget is the officially approved set of documents that define the project objectives, scope, target technical performance specifications, master schedule, key milestone dates, and budget allocations. The baseline concept is similar to the "design freeze" point on the product design side: Once established, changes to the baseline targets can only be made after considered review and approval using established change control procedures. Some baseline changes may fall within the scope of authority of the project manager, especially if they can be handled by allocating time or money from the management reserve accounts, but many will require approval of the project sponsor, project owner, or the customer providing funds for the project.

As with a number of the more formalized project management concepts described in this book, the baseline concept originated with the U.S. Department of Defense (DOD). Kemps describes the concept well:

> Since 1967 the DOD has employed the Cost Schedule Control Systems Criteria (C/SCSC) as a means to ensure that major contractors' internal management systems are sound and can provide government program managers with reliable, objective cost performance information for use in management decision making. The "criteria approach" allows the contractors to adopt the systems and controls of their own choosing, providing those systems can satisfy the criteria. Compliance is determined by government teams which review the systems in operation after contract award.
>
> The C/SCSC require that a contractor establish an integrated cost and schedule baseline plan against which actual performance on the contract can be compared. Performance must be measured as objectively as possible based on positive indicators of physical accomplishment rather than subjective estimates or amounts of money spent. Budget values are assigned to scheduled increments of work to form the performance measurement baseline.
>
> The importance of the baseline is evidenced by the fact that of the 35 C/SCSC Criteria, more than a third (11) are concerned with establishment and maintenance of the baseline. The reason for such emphasis is that establishment of the baseline forces recognition of the entire scope of work and how it will be accomplished. Maintenance of the baseline is intended to avoid uncontrolled changes to baseline budgets and schedules that can result in meaningless performance data.[1]

As described in Chapter 14, the more useful project scheduling software packages show the baseline plan on the schedule charts they produce, and compare the current plan and schedule graphically against the baseline schedule.

Figure 12.4 illustrates the relationships between a number of directives and planning documents and the baseline plan for a large, complex,

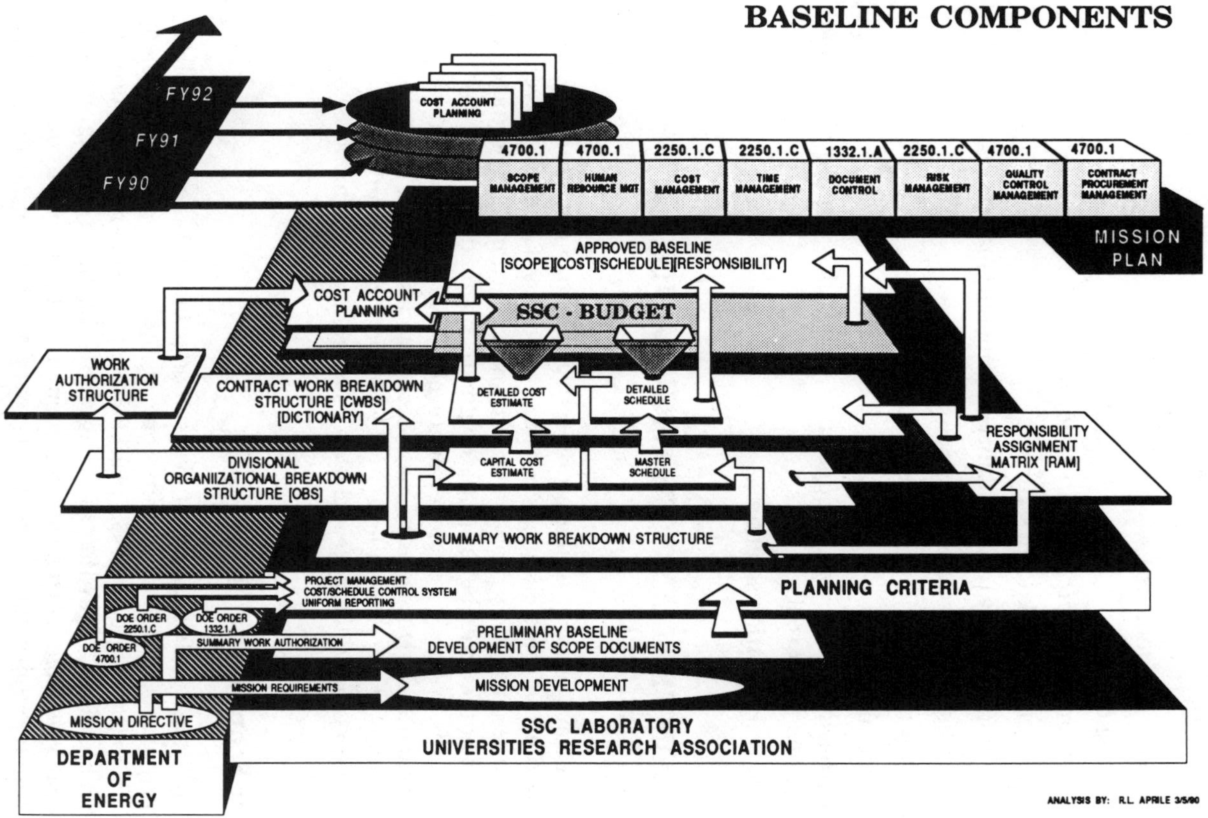

BASELINE COMPONENTS

Figure 12.4 The baseline plan process and components for the Superconducting Super Collider Project. (Source: Superconducting Super Collider Laboratory, Universities Research Association. Used by permission.)

267

high-technology project. Highly visible in the portrayed process of building the baseline plan for this project are many of the planning documents described in Chapter 10, and the work authorization and control documents discussed in this present chapter:

- Project scope documents.
- Planning criteria.
- Project and contract/work breakdown structure.
- Organization breakdown structure.
- Task/responsibility (responsibility assignment) matrix.
- Project master schedule.
- Detailed task schedules and cost estimates.
- Project budget.
- Approved baseline plan: scope, cost, schedule, responsibility, and components thereof.
- Work authorization structure.

12.3 CONTROLLING CHANGES AND PROJECT SCOPE

Projects involve accomplishing an objective that has not been achieved before under the same circumstances or conditions. Changes, therefore, are inevitable since every problem or circumstance cannot be predicted or anticipated at the time the project is originally planned.

Procedures are required to evaluate and control changes in scope, schedule, and cost. Decisions to make a change may be made at the task project leader, at the project level by the project manager in coordination with affected functional project leaders, by higher management, or by the customer.

Project Scope Control

A primary cause of delay and cost overrun on projects is the uncontrolled, frequently unnoticed, increase in scope of the work being performed, compared to the original project plan. People tend to want to produce the best possible result, regardless of the agreed upon objective. Engineers, if not properly managed, often continue to improve their designs beyond the specification requirements. Customer representatives often exert pressure to perform additional work to correct an oversight on their part, respond to changes in requirements, or improve the product beyond the terms of the contract. Continual surveillance and discipline at the task and project levels is required to control the scope of work.

The project manager plays a key role in this effort, by the following:

- Insisting upon well documented task statements of work, schedules, and budgets, with signatures.
- Monitoring results with the functional project leaders to assure that the specifications and contract conditions are being met—no more, no less.
- Monitoring schedule and budget variances with the functional project leaders to identify tasks where the scope of work may have expanded, and initiating appropriate corrective action.
- Insisting upon revised work statements, schedules, and budgets where an increase in scope is required by the customer or for other considerations. (Contract price changes should of course be initiated when justified.)
- Controlling and monitoring direct contact with customer representatives at the working level, and preventing company personnel from making unauthorized agreements that change the scope of work on any task.
- Personally participating in any trade-off agreements made between task managers and with customer representatives.
- Insisting that the contract administrator be fully involved in the above actions.

Task Work Order Control

Since the task work orders are in effect contracts between the project manager and the functional managers and project leaders performing the tasks, and since the task schedules and budgets are a part of the work orders, these documents are vital to overall project control.

Procedures are required to govern the approval and issuance of work orders, or equivalent documents, and for their revision and administrative close-out. It is important that task work orders be

- Released for a given task only when the start date of the task is near. (However, the functional managers would have preliminary copies of the work orders earlier for planning purposes.)
- Used as the basis for evaluation of progress in schedule and cost, determination of cost and schedule variances as described later, and estimation of the cost to complete.

- Updated and revised periodically to reflect actual progress and new knowledge.
- Terminated, and the associated cost accounts closed, immediately upon completion of the task.
- Retained as part of the project file for record purposes and for use in estimating future tasks of a similar nature.

Subcontract and Purchase Order Control

Established procedures should be used to carry out similar control actions related to subcontract and purchase orders. The project contract administrator is responsible for the duties specified in Section 9.5 for all subcontracts. The designation of a purchasing project leader will simplify control of all purchase orders for the project.

Engineering Change Control and Configuration Management

Ineffective control of engineering changes is frequently the cause of delay and cost overruns on projects. To manage this,

- Procedures for controlling engineering changes must be in operation.
- A *Change Control Board* must be established for the project, with voting membership to include the project manager (who may act as chairman in some cases) and key managers and/or functional project leaders from engineering, manufacturing, marketing, purchasing, contract administration, and possibly other areas.
- A *design freeze point* must be established, corresponding to the "baseline" design, and all subsequent design changes rigidly controlled and documented by the Change Control Board, which also determines the point of effectivity of each change (model or serial number to be affected).
- A disciplined procedure must be established and followed for controlling and documenting the functional and physical characteristics of the products being designed and produced. This is referred to as configuration management.

Contract Administration and Control

The contract administrator plays a key role in controlling the project during its execution, as detailed in Section 9.5, and assists in controlling the scope of work.

12.4   SCHEDULE CONTROL

Schedule and cost control must be performed on an integrated basis to achieve effective project control. In addition, technical progress on performance must be measured (see Section 12.8) and correlated to schedules and costs. Concentrated effort is necessary in each of these areas to achieve control of schedules and costs, integrate schedules and costs, and to measure technical progress and correlate to schedules and costs.

Correlation of Schedules, Costs, and Technical Progress

Schedules and costs are correlated at the task or work package level by the task schedule and budget document, as a part of the task work order. The correlations can be summarized in several ways, or a combination of these, for higher level control purposes. Examples include:

1. By elements of the project breakdown structure.
2. By organization (section, department, etc.).
3. By financial ledger account (engineering expense, manufacturing, etc.).
4. By type of cost (labor, material, overhead).

The project manager generally finds the summary by elements of the project breakdown structure the most useful for project control purposes.

Figure 12.5 illustrates the correlation of schedules and costs with the project breakdown structure. The PBS is arranged on the left side, and the work packages or tasks are shown by the bars, in line with their parent level PBS elements, in this case level 3. The project master network plan is partially illustrated, linking the tasks through milestone and other events. Budgets, actuals, and estimates-to-complete and -at-completion are indicated for a few tasks and for the total project. Intermediate level summaries are also made for each level and element. Technical progress is measured by correlation of technical objectives to specific milestone events, as discussed in Section 12.8.

Requirements for Control

Summarizing a number of the points in previous paragraphs, effective control of project schedules and costs requires:

- *Thorough planning* of the work to be performed to complete the project.
- *Good estimating* of the time, labor, and costs.

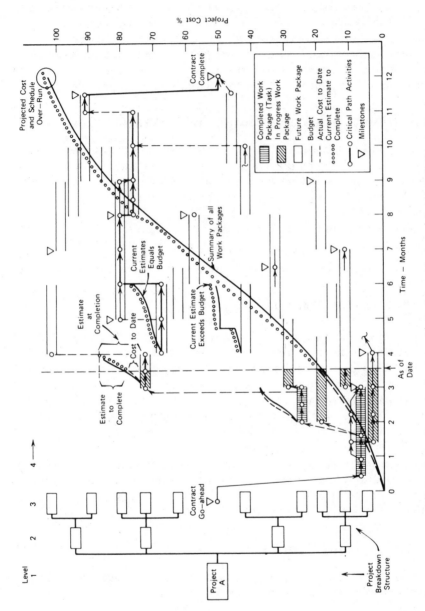

Figure 12.5 Correlation of schedules and costs with the project breakdown structure.

272

- Clear *communication* of scope of required tasks.
- Disciplined *budgeting and authorization* of expenditures.
- *Timely accounting* of physical progress and cost expenditures.
- *Periodic reestimation* of the time and cost to complete remaining work.
- *Frequent, periodic comparison* of actual progress and expenditures to schedules and budgets, both at the time of comparison and at project completion.

This process identifies deviations from plan that indicate the need for controlling action to recover to schedule and/or budget.

Measuring Progress against Schedule

Each functional project or task leader or manager must measure progress against the task schedule on a regular basis. This is usually done at the end of every week, since the weekend is then available for evaluation, special recovery effort, and replanning as needed for the following week.

Whether the task schedule is in the form of a bar chart or network plan, progress measurement is basically the same:

- *Progress to date is recorded*

    Completed activities noted

    Work accomplished on in-progress activities is noted
- *Remaining work is estimated*

    Time to complete in-progress activities is recorded

    Future activities are reestimated or replanned as required
- *Impact on task completion* and other key interface and milestone events is determined.

For in-process work, the emphasis should not be on "percent complete," but rather on *time remaining to complete* each activity and task.

The next step in progress measurement is to look at all tasks within the next higher level element of the project breakdown structure, with emphasis on intertask interfaces. Does a delay in one task affect other tasks? If so, what is the effect? This is where the use of PERT/CPM/PDM network plans provides powerful analysis of each level of summary, and of the total project master schedule.

The objective is to identify the currently critical activities and tasks that must be completed on schedule to meet key commitment dates. In a typical project network plan about 15% of all activities form the critical path. When these are identified, management attention can be concentrated on them to assure on-schedule completion of the project.

Interface Event Control

Interface management is discussed in Chapter 13. On very large, complex projects, formal interface event planning and control procedures are required. On smaller projects, less formal methods can be used. In either case, the project manager should:

- *Establish interface event lists*
    Interface event coordination list
    Approved interface event list
    Interface event revision list.
- *Provide each task manager with lists* of the incoming and outgoing interface events for his task(s).
- Provide each task manager with the *current predicted dates* for all *incoming interface events,* based on overall integrated evaluation of the project master schedule, on a weekly or monthly basis, as appropriate.

When incoming interface events are delayed, the task manager usually must replan the work to attempt to avoid delay of subsequent outgoing interfaces, including task completion. This often incurs added cost, and the project manager analyzes the trade-off between schedule and cost with the concerned functional managers. If necessary, a portion of the project management reserve may be allocated to cover the increased cost to recover the delay.

The earlier a delay of an interface event is made known to the affected managers, the greater the chances are of replanning to overcome the delay without added cost.

12.5   COST CONTROL

Project cost control, like schedule control, is exercised primarily at the task (work package) level by the functional manager or project leader responsible for each task. The project manager's basic responsibility is to monitor costs at the task and higher levels of summary, to identify significant variances between actual expenditures and budgets, and to initiate corrective action so that the total final project cost is equal to or less than the total budget. Cost control for project management purposes consists of:

- *Setting budgets* for specific tasks (Section 10.14).
- *Measuring expenditures* against budget and identifying variances.
- Assuring that *expenditures are proper.*
- Taking *appropriate controlling actions* where budget variances exist.

Measuring Expenditures against Budget

Reports are required weekly, or at a minimum, monthly, to show direct labor, material, and other costs on each in-progress task. These data can be entered directly onto the task schedule and budget document (Figure 10.18), for use by the task leader or manager. Variances will then be apparent, coupled with information on actual progress against schedule.

Monthly actual versus budget reports are required at the total project level, intermediate project breakdown structure levels, and the task level, showing direct costs, burden costs, and total project costs.

Recording and Controlling Commitments

Outstanding commitments that are not yet booked must be included in the actual versus budget costs reports for project control purposes. Procedures must be established to identify and report such commitments as early as possible, even though their final values may change when booked. This would include reconciling the commitments to the actual costs when the bills are paid. Failure to record and control commitments is a frequent cause of cost overruns, especially where large purchases or contracts are required.

Late Charges

Late charges may result from disputes on invoices, correction of administrative errors, late submittal of invoices, and similar actions. Good commitment reporting and control will eliminate many late charges, but some are unavoidable. Any substantial apparent budget underruns should be analyzed carefully before reallocating the funds, to prevent surprise by late, forgotten charges.

Assuring That Expenditures Are Proper

Each task manager must continually assure that expenditures of labor or money charged to a task account number are proper and are for effort actually contributing to that task. There is a continual temptation for functional managers to charge costs for work done on a project where an overrun exists to another task (usually just beginning) that still has an unexpended budget balance. This hides real problems and distorts the cost records for analysis and future bidding purposes.

Persons performing the work may not realize the importance of properly recording their time against the charge (account) number for each task, and as a result simply fill in their time sheets or cards in a careless fashion. The project manager must continually watch for such erroneous charges to his project, whether or not they are intentional. Internal auditors should be asked to verify periodically that established policies and procedures are in fact being followed.

Cost to Complete and at Completion

Cost to complete should be re-estimated monthly on in-progress tasks, and quarterly on all incomplete tasks. This is commonly termed the ETC—estimate to complete.

Cost at completion must be forecast at least monthly at the task, intermediate, and total project levels. Cost at completion is the sum of:

- Cost of all completed tasks.
- Cost to date plus estimate to complete on all in-progress tasks.
- Current estimate to complete each future task.

This is commonly termed the EAC—estimate at completion.

Controlling the Management Reserve

As discussed in Section 10.13, all reallocations of authorized funds for individual tasks must be recorded in the management reserve transaction register. The project manager should approve such transactions together with other approvals as required by established policies.

Causes of Cost Problems

Some of the primary causes of cost problems on projects are:

- Unrealistic, low original estimates, bids, and budgets.
- Management decision to reduce bid price and budgets to meet competitive pressure or offset assumed inflated estimates.
- Uncontrolled, unnoticed increase in scope of work.
- Extrascope work on proposals for change or extensions, or in response to customer or management inquiries.
- Unforeseen technical difficulties.
- Schedule delays that require overtime or other added cost to recover, or charging of idle labor time to the project.
- Inadequate cost budgeting, reporting, and control practices and procedures.

Analysis of cost control experience for specific projects should be made by each company to identify the sources of cost problems and initiate the needed corrective actions.

Project Cost Accounting Problems

If the company cost accounting and reporting system and practices are inadequate, good project cost control will be difficult. Typical problems in this area are:

- Information is not timely.
- Project chart of accounts is not set up to meet project management needs.
- Commitments are not properly recorded and reported.
- Manufacturing costs are difficult to identify until item is shipped, or to allocate and compare to budget.
- Project summaries (using the PBS) are not produced.
- Cost to complete and cost at completion are not handled in the accounting system or reporting procedures.

12.6 INTEGRATED SCHEDULE AND COST CONTROL: THE EARNED VALUE CONCEPT

Experience over the years on many projects in many industries, as reported in the extensive project management literature available today, has shown that trying to control physical progress and schedules separately from costs usually results in ineffective project control. However, achieving integrated schedule and cost control is a complex and demanding task. The planning principles and methods described in Chapter 10 are designed to enable such integrated control. The systematic definition of the project down to the work package or task level provides the key. For each task/work package, the start, completion, and any intermediate interface events are linked directly to the task estimates and budgets, as shown in Chapter 10.

Earned Value

As the term implies, the earned value of a task is the approved budget allocated to perform the task. When the task is complete, the value of the budget has been earned. The concept is simple and powerful, but there are many difficulties in using it effectively, especially on large projects. It is most effective when there are a large number of tasks and they are relatively short in duration compared to the status reporting period, since that would provide a number of task completions for variance analysis. However, breaking down a project into a large number of tasks may create such an administrative workload that it is too cumbersome and costly to maintain the information up to date. If the number of tasks is too few, they will all be of long duration, which erodes the effectiveness of the earned value measurement. With a task of long duration, if the value is not earned until its completion, substantial effort and money will be expended before any measure of progress can be made. If one resorts to estimates of how far along the task is prior to completion, the measurement loses its objectivity. This dilemma has led to various rules that may be dictated or allowed for a given project.

Fleming[2] identifies six of these rules:

1. The 50/50 (or other ratio) technique: used for work packages with durations of two or three reporting time periods; 50% is earned at the start, 50% at completion.
2. The 0/100 technique: best applied for very short work packages. No value is earned when started, none while it is in progress, and 100% is earned on completion.
3. Milestone method: used for longer-duration tasks; earned value is assigned to objective milestones and credited when the milestones are reached.
4. Percent complete: with firm guidelines, can be made to work; recommend setting an upper limit of around 80% for earned value credited based only on estimates of progress, with the balance earned on completion.
5. Equivalent and/or completed units: useful for production tasks; earned value is based on a pre-set value for completion of specified units.
6. Earned standards: requires sophisticated methods to set standards for hours worked.

The use of subjective, unsubstantiated estimates of "percentage complete" of long-duration tasks, or of the total project, is generally unreliable, since these normally show that everything is on schedule and budget—until the last 5 or 10% of the task or project, which will then require an inordinate amount of time and money to complete.

Methods of handling various types of work packages—discrete, level-of-effort, apportioned—have to be tailored to the customer demands and project needs.

C/SCSC Measurement of Schedule and Cost Variances

The key terms in this area introduced in the 1960s with the DOD C/SCSC described earlier are:

- Budgeting cost of work scheduled (BCWS): the budgeted value of one or more tasks for a given time period.

- Budgeted cost of work performed (BCWP): the budgeted value of the work that has been reported complete up to the reporting date; the earned value for that work. Usually obtained from the project and task budgets and work orders.

- Actual cost of work performed (ACWP): the actual cost of the work reported complete up to the reporting date, usually obtained from the cost accounting system.

Earned value and variance analysis can be performed at the task/work package level for each item reported complete, and by summarizing the information using the project breakdown structure, at each level up to the total project.

Schedule Variance

The schedule variance is the difference between the budgeted cost of work performed and the budgeted cost of work scheduled (BCWP minus BCWS). It is a measure of how far ahead or behind schedule the task or project is in monetary terms. The schedule variance can also be approximately interpreted in time units by dividing this number by the rate of expenditure at that time.

Cost Variance

The cost variance is the difference between the budgeted cost of work performed and the actual cost of work performed (BCWP minus ACWP). In other words, it is a comparison of the earned value with the actual cost of the work reported complete.

Variance and Trend Analysis

Either of these variances can be favorable or unfavorable. Careful analysis is required to assure that the information being reported is reasonable and consistent. It is generally useful to set up simple charts to monitor the trends over several reporting periods, rather than jump to conclusions based on one-time reports. Examples of typical charts used in monitoring and progress evaluation are given in Chapter 15. A graphic portrayal of schedule and cost variances is shown in Figure 15.2.

Compliance with the C/SCSC on large military and aerospace programs is very complex. Fleming's book[3] provides a comprehensive description of these criteria and their application, together with the basic governmental C/SCSC documentation and a complete glossary of terms. The Performance Management Association (see Note 1) is a fast-growing nonprofit professional society devoted to this special segment of the project management profession.

There are a number of available computer software packages (discussed in Chapter 14) that will calculate earned value and the variances associated with this concept. One example of a useful chart plotted by one of these software programs is given in Figure 12.6, showing both the bimonthly rates and variances together with the cumulative totals.

12.7   C/SCSC COST/SCHEDULE PERFORMANCE REPORTS

A means of measuring overall cost and schedule performance on the project is required. The basic content and format of the cost/schedule performance reports have been established by the U.S. Department of Defense for this purpose, and are recommended for adoption as appropriate.

Cost Performance Report—Project Breakdown Structure (Figure 12.7)

Entries are made on this report for each summary element in the project breakdown structure. The report can be prepared for any level of summarization in the PBS. Schedule variance information is calculated by the task managers or leaders using their task schedule and budget documents.

Column 12 is the currently approved baseline budget for each item, including the management reserve, as shown in the project budget. It is changed only by formal reprogramming to reflect changes in scope or other similar major actions.

Column 13 is the latest revised estimate for each item, including the management reserve. These entries change when it is necessary to reallocate funds from the management reserve, or from one item or task to another, to cover effort that cannot be accomplished for the original amount budgeted. Where work can be accomplished for less than the BCWS, the management reserve will then be increased. All such changes in allocation are recorded in the Management Reserve Transaction Register. The same information can be summarized by functional organization categories, and labor hours can also be shown instead of dollars.

Cost Performance Report—Program Analysis

This narrative report is prepared by the program or project manager to: identify *significant variances* (positive and negative) shown in the cost/

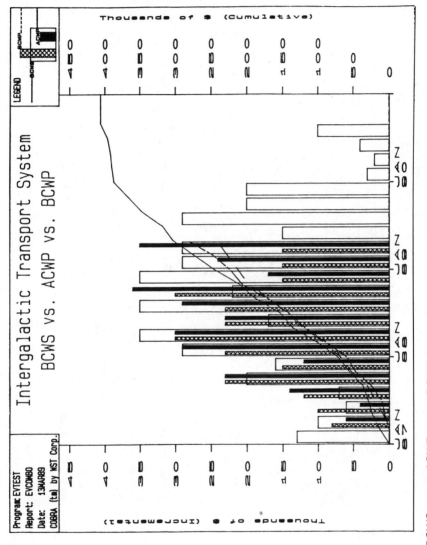

Figure 12.6   BCWS versus ACWP versus BCWP, bi-monthly and cumulative. (Plotted by the Cobra™ software package from Welcom Software Technology. Used by permission.)

COST PERFORMANCE REPORT - WORK BREAKDOWN STRUCTURE

| CONTRACTOR: VST Aerospace Corp. | CONTRACT TYPE/NO: CPFF 09-88-10 | PROJECT NAME/NO: SDI-776 | REPORT PERIOD: 01OCT88-31OCT88 | SIGNATURE: |
| LOCATION: Houston, Texas | | | | TITLE / DATE: |

| QUANTITY 1 | NEGOTIATED COST 260092 | EST. COST AUTHORIZED UNPRICED WORK 0 | TGT. PROFIT/FEE % 26009/10% | TGT. PRICE 286101 | EST. PRICE 303397 | SHARE RATIO N/A | ESTIMATED CONTRACT CEILING N/A |

| | CURRENT PERIOD | | | | | CUMULATIVE TO DATE | | | | | AT COMPLETION | | |
| | BUDGETED COST | | ACTUAL COST WORK PERFORMED | VARIANCE | | BUDGETED COST | | ACTUAL COST WORK PERFORMED | VARIANCE | | | | |
(1) WORK BREAKDOWN STRUCTURE ITEM	WORK SHEDULED (2)	WORK PERFORMED (3)	(4)	SCHEDULE (5)	COST (6)	WORK SCHEDULED (7)	WORK PERFORMED (8)	(9)	SCHEDULE (10)	COST (11)	BUDGETED (12)	LATEST REVISED ESTIMATE (13)	VARIANCE (14)
Primary Vehicle	6351.00	6173.00	6664.00	(178.00)	(491.00)	37249.00	33493.00	40455.00	(3756.00)	(6962.00)	108798.00	131413.00	(22615.00)
Secondary Vehicle	3219.00	3310.00	3080.00	91.00	230.00	19655.00	22080.00	21087.00	2425.00	993.00	59876.00	57183.00	2693.00
Training	15.00	15.00	16.00	.00	(1.00)	103.00	104.00	100.00	1.00	4.00	543.00	522.00	21.00
Peculiar Support Eq.	526.00	524.00	532.00	(2.00)	(8.00)	1534.00	1434.00	1633.00	(100.00)	(199.00)	8654.00	9855.00	(1201.00)
Systems Test & Eval.	231.00	230.00	232.00	(1.00)	(2.00)	765.00	722.00	801.00	(43.00)	(79.00)	4480.00	4970.00	(490.00)
Project Management	65.00	62.00	67.00	(3.00)	(5.00)	105.00	104.00	110.00	(1.00)	(6.00)	339.00	359.00	(20.00)
Data	76.00	70.00	80.00	(6.00)	(10.00)	154.00	143.00	134.00	(11.00)	9.00	975.00	914.00	61.00
Operational/Site Act	245.00	300.00	330.00	55.00	(30.00)	997.00	970.00	1002.00	(27.00)	(32.00)	9875.00	10201.00	(326.00)
Common Support Equip	196.00	207.00	188.00	11.00	19.00	1003.00	1111.00	1045.00	108.00	66.00	12087.00	11369.00	718.00
Industrial Facil.	332.00	180.00	200.00	(152.00)	(20.00)	2543.00	3333.00	3000.00	790.00	333.00	54367.00	48935.00	5432.00
Initial Spares	10.00	12.00	9.00	2.00	3.00	45.00	46.00	44.00	1.00	2.00	98.00	94.00	4.00
GEN. AND ADMIN.	1690	1662	1710	(27)	(47)	9623	9531	10412	(92)	(801)	39014	41372	(2358)
UNDIST. BUDGET													
SUBTOTAL	11266	11083	11398	(183)	(315)	64153	63540	69411	(613)	(5871)	260092	275815	(15723)
MANAGEMENT RESRV.													
TOTAL	12956	12745	13108	(210)	(362)	73776	73071	79823	(705)	(6752)	299106	317187	(18081)

RECONCILIATION TO CONTRACT BUDGET BASELINE

| VARIANCE ADJSTMT. | | | | | | | | | | | | | |
| TOTAL CONTR. VAR. | | | | | | | | | | | | | |

All $ amounts in thousands

Figure 12.7 Cost Performance Report—Work Breakdown Structure. (Produced by the Open Plan™ software package from Welcom Software Technology. Used by permission.)

schedule performance reports and other significant problems. It also describes the *underlying causes* of the variances or problems, states what *corrective actions* have been initiated, if any, and indicates when the problem is expected to be corrected or resolved.

12.8   TECHNICAL PERFORMANCE MEASUREMENT

Technical performance measurement (TPM) is the continuing prediction or demonstration of the degree of anticipated or actual achievement of technical objectives. It includes an analysis of any differences among the *achievement to date, current estimate,* and the *specification requirement.* It is generally the most difficult of the three basic areas of project planning and control: schedule, cost, and technical.

Achievement to date is the value of a specified technical parameter estimated or measured in a particular test and/or analysis.

Current estimate is the value of a specified technical parameter to be achieved at the end of a project or contract if the current plan is followed.

Purpose of Technical Performance Measurement (TPM)

The purpose of TPM is to:

- Provide visibility of actual versus planned technical performance for correlation to schedules and costs.
- Provide early detection or prediction of technical problems that require management attention.
- Support assessment of the impact on the project of proposed change alternatives.

Relationship to Cost/Schedule Performance Measurement

Schedules, costs, and technical results are always interrelated. Cost/schedule performance measurement may reveal problems in the technical area, and technical problems revealed by TPM can surface inadequacies of time or money. However, the cost/schedule measurement methods previously described basically assume that the technical effort and results are adequate for project success. A disciplined approach to TPM is therefore necessary to avoid the problem of completing a project on time and within budget but with a product that is not acceptable.

TPM assessment points should be planned to coincide with the planned completion of significant design and development tasks, or aggregation of tasks. This will facilitate the verification of the results achieved in the completed task in terms of its technical requirements.

Technical Parameters

The parameters to be tracked and reported must be:

- Key indicators of project success.
- Interrelated by construction of tiered dependency trees similar to the specification tree.
- Correlated to an element of the project breakdown structure.

For each technical parameter to be tracked, the following data, as appropriate, should be established during the planning of the related task:

- Specification requirement.
- Time-phased planned value profile with a tolerance band (illustrated in Figure 12.8). The profile must represent the expected growth of the parameter, and the boundaries of the tolerance band must represent the inaccuracies at the time of the estimation, and also indicate the region of budget and schedule within which the required specification is expected to be achieved.
- Project interface or milestone events significantly related to the achievement analysis or demonstration.
- Conditions of measurement (type of test, simulation, analysis, etc.).

Conducting Technical Performance Measurement

As the design and development activity progresses, achievement to date is tracked continually on the planned value profile for each technical performance parameter. In case the *achievement to date* value falls outside the tolerance band a new profile or current estimate is developed immediately.

The current estimate is determined from the achievement to date and the remaining schedule and budget. Any variation outside the tolerance band is analyzed to determine the causes and assess the impact on higher level parameters, interface requirements, and system or product cost effectiveness.

For technical performance deficiencies, alternate recovery plans are developed with cost, schedule, and technical performance implications fully explored. For performance in excess of requirements, opportunities for reallocation of requirements and resources are assessed.

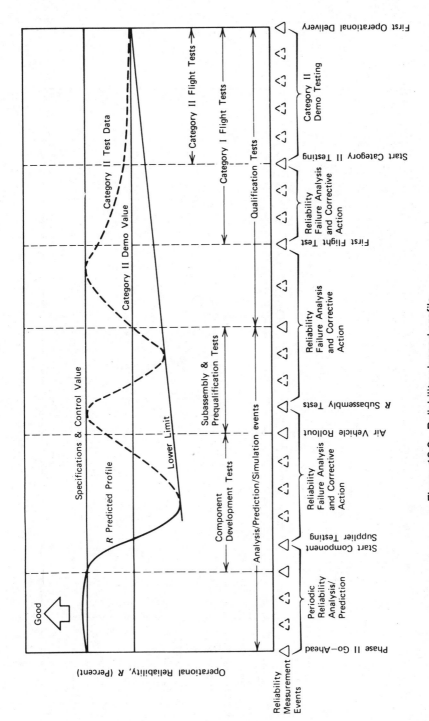

Figure 12.8 Reliability planned profile.

285

PROGRAM _____
CONTRACT NO. _____
ITEM _____
SPEC NO. _____
WBS NO. _____

Data Item
S–C584–001B

TABLE I
TECHNICAL PERFORMANCE MEASUREMENT REPORT

SUPPLIER NAME _____
RESP. ENGINEER _____
PROJECT MANAGER _____
DATE ISSUED _____
REVISED _____

*PARAMETER REQUIREMENTS

*PARAMETER	SPEC PARAGRAPH REFERENCE	PARAM UNITS	SPEC VALUE	CONTROL VALUE	EVENT NO. / *DATE	EVENT NO. / DATE	EVENT NO. / DATE	EVENT NO. / DATE	EVENT NO. / DATE	EVENT NO. / DATE	EVENT NO. / DATE	EVENT NO. / DATE	EVENT NO. / DATE
Dection	3.1.1.3.2.4												
Antenna Pattern	3.1.1.3.5.2.1												
Missile Illumination	3.1.1.3.2.4.9												
Range Accuracy	3.1.1.3.2.9.1												
Angle Accuracy	3.1.1.3.2												
Range Rate Accuracy	3.1.1.3.2.												
Angle Track Rate	3.1.1.3.2.4.4												
Mov–up–Date	3.1.1.3.2.4.1.1												
Power Input	3.1.1.1.1.3												
Reaction Time	3.1.1.4												
Reliability	3.1.2.1.1												
Maintainability	3.1.2.2												
Weight	3.2.2												
Volume	3.2.2												

Figure 12.9 Technical performance measurement report.

Technical Performance Measurement Report

For a given subsystem or element in the project, a summary technical performance measurement report is prepared, as illustrated by Figure 12.9. Each parameter is listed, with the specification reference and the parameter units, specified value and control value. As key events are reached the current estimate is recorded for each parameter.

13

‹ ›

Project Interface
Management[1]

13.1 WHY PROJECT INTERFACE MANAGEMENT

Planning and executing projects is challenging work for all persons involved: the project manager; the project planning and control staff (if any);
the contributing functional managers, project leaders, and specialists (including outside contractors, consultants, vendors, and others); and the
senior managers to whom these people report.

In spite of decades of experience in positioning the project manager
within our organizations, and the availability of and experience in using
powerful planning and control systems and procedures in project management, there are many areas that could be improved. Often there is too
much conflict, too little acceptance of the role of the project manager, too
little real use of the advanced planning and control systems that are available today, and projects that are not planned and controlled as effectively
as they can and should be.

One approach to attacking these areas of need in project management
can be found in the practice of *project interface management.* Experience in
the application of this approach shows that it can lead to significant improvements in the following respects:

- Better definition of roles and responsibilities

  Improved understanding of the role of the project manager and
  better acceptance of the need for that role.

Clarification of the roles and responsibilities of the functional contributors to the project.

Reduction in conflicts between the project manager and the functional contributors, as well as among the individual functional managers, project leaders, and specialists.

- Project planning and control improvements

Provision of good, logical linkage points between the levels in the schedule hierarchy, and between subprojects and subnetwork plans.

Better identification of the proper degree of detail at each schedule level.

Wider acceptance and use of project planning and control systems.

- Teamwork and team building improvements

Improved teamwork through clear identification of the points of interaction between functional tasks.

More effective team building through joint identification of the key project interface points.

13.2   THE CONCEPT: THE PROJECT MANAGER AS THE PROJECT INTERFACE MANAGER

The basic concept of interface management is that the project manager plans, schedules, and controls—in a word, manages—the key interface events on the project, while the responsible functional project leaders manage the tasks or work between these interface events. This is no more than recognition, systemization, and formalization of how the project manager and functional contributors should divide their responsibilities, and how they should work together on any given project. The project manger must plan, schedule, and control the project interfaces in close cooperation with the contributing functional project leaders.

Various Meanings of "Interface"

The word "interface" can mean different things to different people. Often you see "interface agreements" as part of the project procedures, especially on large engineering/construction projects. These agreements establish the ground rules governing the relationships between the owner, the project manager, the architect engineer firm, and major contractors. In other words, they deal with the ongoing organizational relationships (or interfaces) between the major parties involved in a project.

Another meaning relates to the interaction between phases of a project, and there are many references in the literature to the engineering/construction interface, for example. This is closer to the meaning of "project interface management" used in this book, but not exactly the same thing.

Project Interface Management

While the management of these kinds of "interfaces" could be considered a form of "interface management," this term refers more specifically to managing the specific project interfaces, as defined next. As used here, it is therefore somewhat different from managing on-going interfaces between organizations in a general sense, or generally managing the interface between engineering and construction, or other *interphase* management activities.

13.3 PROJECT INTERFACE MANAGEMENT IN ACTION

For a specific project, implementation of project interface management involves these activities during project start-up, execution, and closeout:

- Project or new phase start-up planning

  During the initiation of the project, or of one of its major phases, the key project interface events are identified and described, planned, scheduled; outgoing and incoming responsibilities are assigned to specific individuals.

- Project or phase execution

  During project execution, the project interface events are monitored and controlled as part of the on-going procedures used to manage the project.

- Project closeout

  During the close-out phase, the project interface events that occur at the end of the project form important elements of the closeout checklists, and assist the project and functional managers in assuring that all the loose ends of the project are tied up so that the project can be completed cleanly.

- Identifying project interfaces through input-output analysis
- The key project team members, once they have studied and understood the project objectives and scope, can usually identify most if not all of the key interfaces by performing an input-output analysis of the tasks they will be responsible for on the project.

This analysis simply requires each task leader to think through two questions:

1. What inputs do we need (information, resources, approvals, other) to initiate and then complete this task, and who will provide these?

    This will identify the *incoming* interfaces for that task.

2. What intermediate and final outputs will we generate in the performance of this task, and who should receive these?

    This will identify the *outgoing* interfaces related to the task.

An effective practice is to ask the functional managers, project leaders, or task leaders to prepare a memorandum for each major task they are responsible for on the project, listing the identified incoming and outgoing interfaces, with the expected sources and recipients for each. These memos should then be distributed to all affected project team members. It is sometimes surprising, at least to some team members, to see what others are expecting of them. In other cases, a task leader may not be aware that a particular team member needs to receive the output from a given outgoing interface. By sharing these memos with the other team members, each person has the opportunity to verify who is expecting what input from whom, and who is planning to give what output to whom. Better teamwork, improved communications, and fewer omissions and mistakes are the result.

A useful form for task input-output analysis from Tuman,[2] enabling a more rigorous approach to identifying key interface points, is given in Figure 13.1.

This approach also helps to position the project manager as the project interface manager. In this role, she or he is seen by all members of the project team as fulfilling a vital function that will aid the entire team in achieving success.

13.4 PRODUCT AND PROJECT INTERFACES

It is useful to differentiate between product and product interfaces. *Product* interfaces deal specifically with the things being created by the project activities: the intermediate and final results or products of the project. *Project* interfaces deal with the process of creating these products.

Product interfaces fall into two categories: (1) *Performance interfaces,* which exist between product subsystems or components; and (2) *Physical interfaces,* which exist between interconnecting parts of the product.

Note that the products or results of the project can be hardware, software, services, new consumable products, physical facilities, documents, and information. Performance and physical interfaces exist in all of these types of products.

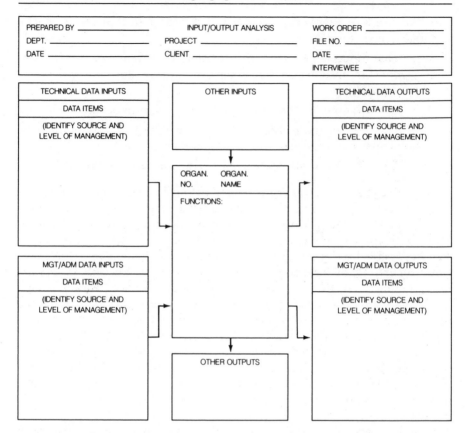

Figure 13.1   Input/output chart. (Source: Tuman, John Jr., "Development and Implementation of Project Management Systems," *Project Management Handbook,* 2nd Ed., David I. Cleland and William R. King (Eds.) (New York: Van Nostrand Reinhold, 1988) p. 666. Used by permission.)

For engineered products, procedures for managing the product design, quality assurance, and product configuration will provide the required management of the product interfaces, both performance and physical. For other types of products equivalent procedures must be provided.

Types of Project Interfaces

Six categories of project interfaces can be identified, although the lines of distinction between some of these can be rather hazy:

1. *Change of responsibility:*   One task is completed and the task product is handed over to another team member or organization for further work. A large percentage of project interfaces are of this type.

Example: Engineering completes a specification, then Purchasing initiates procurement of the item specified. Transmittal of the specification from Engineering to Purchasing is an interface event (outgoing for Engineering, incoming for Purchasing). This example is also an information interface.

2. *Result of Action:* Results from one task are required before another task can begin.

Example: Foundations must be completed by the concrete contractor before the process equipment can be set in place by the equipment erection contractor. This could also be considered a change of responsibility interface.

3. *Management:* Key decisions, approvals, and other management actions affecting other project interfaces, specific tasks, or the overall project.

Example: Senior manager approval of the contract award for development of a software system needed for the project.

4. *Customer:* Actions similar to management interfaces, but involving the customer or client.

Example: Customer approval of the conceptual system design in a software development project.

5. *Information:* Information or data developed in one task and needed by one or more other tasks.

Example: Information on the soil conditions obtained during site investigations by the geologic survey engineering consultant is needed by the civil engineers designing the foundations.

6. *Material:* Equipment, supplies, facilities, or other physical items must be available at a specific location for work to proceed.

Example: A mobile crane needed to hoist a major piece of equipment into place must be moved from another area of the project.

13.5   PROJECT INTERFACE EVENTS

All of these six types of project interfaces can be represented as *project interface events.* Events are points in time associated with specific dates (predicted, scheduled, or actual) that indicate when an action has taken place. Many project interface events represent doing something to a product interface, as for example, "Product XX Specification Released," or "System YY Design Approved." The specification released by engineering contains performance and physical information on a specific part of the product being created, enabling the purchasing department to procure the item in question so that it will perform and fit within the total project result and schedule.

Interface events are important elements of any comprehensive project plan. The most important ones must be included in the Project Master Plan and Schedule, and all key interface events must be included in the integrated project network plan. Interface events provide the means for integrating sub-nets at the second, third, or lower level planning tiers into the overall project plan and schedule. Many management milestone events are also interface events.

13.6 THE FIVE STEPS OF PROJECT INTERFACE MANAGEMENT

Project interface management consists of five steps:

- Identifying.
- Documenting.
- Scheduling.
- Communicating.
- Monitoring and Controlling.

Identification

The first step in interface management is to identify the key project interface events. Clear, unambiguous event identification is required. An event occurs at a point in time and is different from an activity or task, which consumes time. An event signifies the start or completion of one or more activities or tasks. Events must be identified and defined so that they are recognizable when they occur. Their identifying description should relate each event to an element of the project/work breakdown structure. Each interface event is "outgoing" for the originator (usually only one) and "incoming" for each receiver (of which there may be several).

The checklist presented in Chapter 10 (Figure 10.11) should be helpful to clearly and unambiguously identify such events.

Documenting Interface Events

On smaller projects, inclusion of well-identified interface events in the project plans and schedules is usually all that is required. However, on large, complex projects formal procedures are required to document and control interface events. These usually provide for three interface event lists:

1. *The Interface Event Coordination List,* covering new events not yet on the other lists.

2. *The Approved Interface Event List,* which includes all such events which have been coordinated with the affected organizations (originators and receivers) and approved by the project manager.

3. *The Interface Event Revision List,* which includes revisions made to the Approved Interface Event List during the past month (or reporting period).

When such formal procedures are required, due to the size or complexity of the project, number of organizations involved, or geographic dispersal of project contributors, the following basic information is provided:

- Codes: An event code number consistent with the network planning procedures in use on the project, plus an identifier indicating that this is an interface event.
- Description of the event.
- Organizations affected:
    Originator
    Receivers(s)
- Project elements and tasks within the breakdown structure related to the event, and the subnetworks in which it appears.

Scheduling Interface Events

The third step in interface management is to develop a scheduled date for each interface event, reflecting the current integrated project master schedule.

The top-level master project network must include the most important interface events, together with other milestone events of interest to top management. Initial estimates of the time required between the major interface events are made by the persons responsible for each outgoing interface event, and the master project network plan is then analyzed and revised until the key target dates appear to be achievable. At this stage, the second level, more detailed subnets are developed, incorporating the pertinent interface events, and adding detail for the functional tasks involved in each subnet. After each responsible functional manager approves his or her subnets, the results are incorporated into the integrated project network. The subnets can either be integrated into the project network, or the durations between interface events can be entered into the project network on a summary basis. On large projects, integration of all lower level subnets, with all their details, into one integrated project network may prove to be impractical, due to the large volume of information that would have to be updated.

It is not mandatory for lower level schedules to be in the form of network plans, depending on the complexity of the specific tasks involved. In many cases, functional tasks between interface events can be planned and controlled effectively using bar charts, process sheets, or checklists. However, the project manager must assure that the tasks are adequately planned and scheduled, so that there is a reasonable assurance that the future interface event dates that have been promised will, in fact, be met. Weekly or monthly revisions of the estimated time to complete each task (not a percent complete estimate) are required.

On smaller projects, all that may be required is an overall project master schedule that incorporates the key interface events.

Communicating With Interface Events

Communication between members of the project team, with the customer and with upper management can be enhanced through the proper use of interface events. By using the interface event lists, with clear identification of each event, omissions, errors, and confusion can be avoided. Properly coded events lend themselves to accurate reference by electronic mail, telex, telefax, or telephone, for discussion, changes in planning, conflict resolution, and progress reporting.

Affected interfaces should be included in task work orders, contracts, subcontracts and purchase orders, with appropriate language to assure that they are properly planned, scheduled, monitored, and controlled.

Monitoring and Control Through Interface Events

Interface events should be emphasized in normal project monitoring and control procedures. Progress reporting should require a statement of the estimated time remaining to reach each future interface, if their previously predicted completion dates have changed.

If all interface events are well controlled, the project will also be well-controlled. Interface event control is achieved through the procedures used to add events to the approved list, to revise them, and through the schedule review and control procedures. Control of the project is achieved jointly by the project and functional managers by:

- Controlling the interface events.
- Work authorization and control procedures and practices.
- Project directives.
- Project evaluation and review procedures.

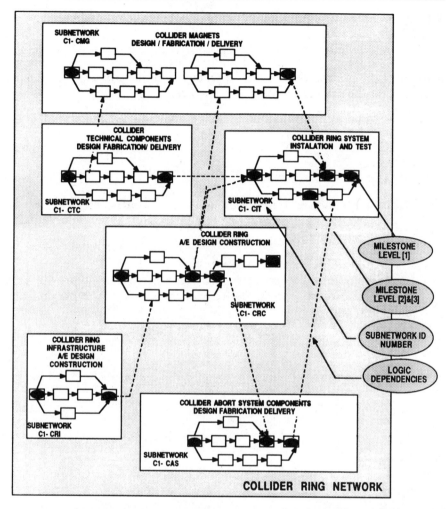

Figure 13.2   Sample diagram showing subnetworks linked through interface events.
(Source: Superconducting Supercollider Laboratory, Universities Research Association. Used by permission.)

Integrating Subnetworks Through Interface Events: An Example

Figure 13.2 shows how detailed subnetwork plans can be integrated through the prior identification of interface and milestone events at several levels of the project breakdown structure.

13.7   CONCLUSION

Good project interface management practices will:

- Clarify roles and responsibilities, and reduce conflicts.
- Define who will provide what to each project team member, enhancing teamwork and communications.
- Improve project planning, scheduling, and control.
- Increase the effectiveness of the project team, thereby improving the chances of project success: Delivery of the specified results on time and within budget.

14

◀ ▶

Project Management
Information Systems

14.1 DEFINING A PROJECT MANAGEMENT INFORMATION SYSTEM (PMIS)

There are many ways to define and depict a project management information system (PMIS). Cleland[1] has defined an overall project management system consisting of five subsystems (planning, information, control, human, and facilitative organizational subsystems) and two additional elements (techniques and methodologies, and cultural ambiance). Tuman[2] presents detailed descriptions and analyses of project management information and control systems from several perspectives, reflecting his long experience in developing and implementing computer-based systems for project planning and control.

Tuman defines a project management and control system of broad scope, as shown in Figure 14.1, including both technical and risk information and control systems, in addition to a project information and control system. For the purposes of this book, a more limited definition of a PMIS is used, focusing primarily on the project as a process, and less on the products of the project. The approach taken here is rather pragmatic: a PMIS is defined as the

- *Documents* (containers of information) and
- *Procedures* for document preparation, maintenance, preservation, and utilization,

299

that are used for creating, planning, and executing projects within a given organization.

PMIS Documents

Many documents for both product and project planning and control have been identified in Chapters 10 and 12, and others are identified in Chapter 15 that are used for evaluating progress, directing, and reporting. The key documents that contain the essential information to be handled in a PMIS as defined here are summarized in Figure 14.2. However, there usually will be a number of additional subordinate documents and procedures; for example, the project and task budgets are dependent on estimates for labor, materials, travel, services, and other expenditures. There will be specific estimating procedures and documents for developing and approving these budgets, reflecting the industry and nature of the project at hand.

Technical Information and Control System	Project Information and Control System	Risk Information and Control System
Engineering Management Module	Work Breakdown Structure Module	Planning Assurance Module
Procurement Management Module	Planning and Scheduling Module	Quality Assurance Module
Construction/Production Management Module	Cost Management Module	Reliability Module
Test Management Module	—Cost Estimating	Maintainability Module
Configuration Management Module	—Cost Est. Support	Safety Assurance Module
	—Craft and Crew	
	—Unit Material	
	—Unit Manhours	
	—Source Document	
	—Cost Control	
	—Cost Projection	
	Accounting Module	
	Data Entry Module	
	On-Line Query Module	

Figure 14.1   Tuman's Project Management Information and Control System. (Source: Tuman, John, Jr., "Development and Implementation of Project Management Systems," *Project Management Handbook,* 2nd Ed., David I. Cleland and William R. King (Eds). (New York: 1988) Van Nostrand Reinhold, chapter 27, p. 673. Used by permission.)

Planning (Ch. 10)	Authorizing (Ch. 12)	Controlling (Ch. 12 and 15)	Reporting (Ch. 15)
Project Summary Plan	Master Contract Release	Management Reserve	Monthly Progress Reports
Project Breakdown Structure (PBS)	Project Release	Transaction Register	—Narrative
Task Responsibility Matrix		Cost Expenditure Reports	—Project Master Schedule
Project Master Schedule	Subcontracts and Purchase Orders	Updated planning and authorizing documents, comparing actuals with budgets and schedules	—Cost Performance Reports
Integrated Project Network Plan			Management Reviews of Critical Projects:
Project Interface and Milestone Event List		—Project Master Schedule	—Major Project Identification Data
Project Budget		—Milestone Charts	—Summary Status Reports
Project Funding Plan		—Other	—Above Reports as required
Project Chart of Accounts		Cost Performance Reports	
Task Statements of Work	Task Work Orders	Schedule Variance Reports	
Task Schedules		Earned Value and Cost Variance Reports	
Task Budgets		Technical Performance Measurement Reports	
Detailed Network Plans		Milestone Slip Charts	
Technical Performance Planned Value Profiles and Milestones		Trend Analysis Charts	
		Task Estimates to Complete (ETC) and Estimates at Completion (EAC)	
		Action Item Lists from Project Review Meetings	

Figure 14.2   Summary of documents for project planning, authorizing, controlling, and reporting.

PMIS Procedures

Any PMIS worthy of the name will include procedures for developing the information contained in the identified documents, and for managing and using these documents. Most organizations that have been managing projects will have such procedures in place. For large or unusual projects, one of the first tasks of the project manager is to select and tailor these procedures to make them project specific. In the typical multiproject situation, as discussed in Chapter 8, the same procedures should be used for all projects, although there may be exceptions for special cases.

14.2 COMPUTER-SUPPORTED PROJECT MANAGEMENT INFORMATION SYSTEMS

Since the coming of the computer age, beginning in the middle of the twentieth century, and concurrent with the introduction in the late 1950s of planning methods such as PERT/CPM/PDM network planning and scheduling, software packages for project planning, scheduling, and control have been important tools for project management. Unfortunately, there has been a tendency in the project management field to equate a PMIS with these computer software packages, and even to equate such packages with project management as a discipline. As a cursory review of Figure 14.2 will indicate, there is more to a PMIS than even the most sophisticated computer software package or system. The project file, discussed in Chapter 10, is simply an orderly collection of these documents, plus related correspondence and other documents of record.

However, computers and software packages are important elements of a modern PMIS. Automation of the handling of the large amounts of complex, interrelated information required to manage a large project, or to manage multiprojects in a dynamic, high-technology environment, has made possible the definition and effective use of integrated project management information systems.

Electronic Creation, Storage, Manipulation, and Retrieval of PMIS Documents

All of the PMIS documents listed in Figure 14.2 can be handled electronically. If each document is recognized as one or more files of information, it becomes apparent that computer files can be established containing the information pertinent to each document. Such files can readily be created, updated, manipulated, and stored; and reports, charts, and graphs in a wide variety of formats can be produced. Some of the procedures comprising the PMIS can be captured in the computer software; other procedures, particularly those dealing with *how* the project manager and project team members are to utilize the information contained in the documents and reports, must be developed and maintained *outside* the computer systems.

The basic computer software packages that are required to automate all of these PMIS documents include word processors, spread sheet packages, database managers, action planners and schedulers (usually using PERT/CPM/PDM network-based methods), resource planning and control packages (usually combined with the scheduling packages), risk analysis packages, documentation control packages, and others, together with various report and graphics output generators. The computer software generally referred to as "project management" packages in most cases are action and resource planning, scheduling, monitoring, and control systems. The "project management" name for these materials is not very accurate, since they handle only a portion of the overall PMIS information.

Mainframes to Minis to Microcomputers

In the 1960s, only mainframe computers were available for PMIS support. They were, and remain, expensive and difficult to use (although when linked with microcomputers their ease of use is greatly increased). Minicomputers appeared during the 1970s, with project planning and scheduling software support, and reduced the cost somewhat, plus making it possible to have computers on site when economically justified. During this period, users who did not have these computers available to them within their organizations could use them on a time-sharing or service bureau basis. Specialized consulting/service bureau companies emerged to satisfy these users' needs.

Beginning with the introduction of the first microcomputer packages for project planning and scheduling in the early 1980s, a revolution has occurred in computer-supported PMIS. As Levine wrote in his 1986 book:

> It was nearly thirty years ago that lumbering mainframes were put into service to support project management. For most of those years access to computerized project management was reserved for large organizations that had management information systems operations, an army of dedicated project control specialists, and barrels of money to spend on hardware and software. Due to developments in computer technology during the first half of this decade [the 1980s], the benefits of computerized project management are now available to the rest of us. In the past few years, the world of automation has been turned inside out by the fantastic success of the microcomputer and its acceptance throughout the business community. Now, with a minimum investment, and bypassing the MIS bureaucracies, anyone can use computers in the business place. Who would have believed just a few years ago that we would have this abundance of project management software available for the casual, as well as the serious, user, much of it at

enticingly low prices. The microcomputer has given us access to sophisticated programs that until recently were the private domain of the information systems gurus.[3]

Growth and Major Trends in the Project Management Software Market

Figure 14.3 shows how the project action and resource planning and scheduling software market has grown since 1965 and indicates the dramatic increase in the acceptance of micro/personal computers in this area of application. For many situations, however, as with large military or aerospace programs and projects, mainframes or minicomputers are still the only viable alternative.

In addition to this proliferation of ever more powerful microcomputers and project-related software packages, several other major changes have occurred in the project management software arena in the past few years. These changes include:

- A continuing shift from pure project scheduling to scheduling integrated with resource management.
- A shift from focusing on large, single projects to handling multiple projects in a common database.
- More automated linkage of project planning, scheduling, and resource management systems with other information and control systems, such as:

    Cost estimating

    Labor reporting of work hours

    Cost accounting

    Document control (purchase orders, contracts, etc.)

    Production control, and so on.

- Wider use of relational database architecture in the project management software.
- The decline and virtual disappearance of time sharing and service bureau mini and mainframe computer data processing for project planning and scheduling.
- Major and continuing improvements in user interaction with the systems:

    On-screen menus

    On-line, interactive, quick response to planning changes

    Graphic user interfaces (GUI) (Apple Macintosh in 1984, Microsoft Windows in 1989).

Number of Project Management
Software Packages Available

Computer Type	1964	1980	1982	1986
Mainframe	60[1]	30[2]	32[3]	49[4]
Mini		10[2]	29[3]	118[4]
Micro/Personal		4[5]	17[4]	166[4]

Market Size, Composition, and Growth Rate[5]
North America

	1985		1990		
	$MM	Share	$MM	Share	CAGR*
Large Systems (Minis and Mainframes)	100	66%	125	29%	4%/year
Micro/Personal					
High-End	11	7%	69	16%	
Low-End	41	27%	237	55%	
Subtotal	55	34%	306	71%	33%/year
Total Market	155	100	431	100	19%/year

*CAGR: Compound Annual Growth Rate.

[1] Joseph J. Moder, and Cecil R. Phillips, *Project Management With PERT and CPM* (New York: Reinhold, 1964) p. 254.

[2] *Project Management Software Survey,* Project Management Institute, Drexel Hill, PA, 1980.

[3] *Project Management Software Survey,* Project Management Institute, Drexel Hill, PA, 1982.

[4] *1986 Buyer's Guide to Project Management Software,* New Issues, Inc.

[5] *Microcomputer Project Management Software Comparison Report* (San Rafael, CA: 1 Soft Decision, Inc., 1986/7) p. 1.1.

[6] *Computerworld,* Dec. 8, 1986, p. 56.

Figure 14.3   Size and Growth of the Project Management Software Market, 1964 to 1990.

- Continual improvements in graphic outputs of the systems: time-scaled network plans, bar charts, and combinations of these (bar-nets, or Gantt charts with the interfacing interdependencies graphically portrayed); time-scaled histograms for people and money, showing both rates of usage and cumulative amounts; project master schedules, with milestones and interface events; project/work breakdown and organization breakdown structures; plus other typical graphic representations of business information (pie charts, bar charts, three-dimensional charts, and so on).

Some microcomputer systems give the user an electronic drafting table, where network action plans can be created by clicking a mouse, instantly scheduled, quickly resource loaded, and revised as required by simply grabbing a bar on the screen and stretching or shrinking it to change its duration.

While it is not possible in a book of this size to illustrate effectively the power, capabilities, and graphic outputs of these systems, three examples are given in Figures 14.4, 14.5, and 14.6. Figure 14.4 shows a small portion of a typical network plan in the precedence diagram format. Figure 14.5 illustrates a logic barchart or bar-net, combining the classic Gantt chart format with the sequential dependencies shown in the network. This example has plotted the work for each activity using the early or expected dates, and graphically indicates the amount of float or slack time available for each. Those activities with no float are termed critical: no delay can be absorbed without delaying the end of the project. These are shown in red on the original of this chart. Figure 14.6 is an interesting illustration of the "rolling wave" concept of project planning, in which the near term work is planned and scheduling in more detail than the work beyond a point a few weeks or months ahead of the time now date. A more detailed one-month window is shown on the left side of the chart, and a condensed nine-month window on the right side.

Additionally, a few of the illustrations used in other chapters of this book are reports produced by the packages identified in each figure. The use of these examples is not intended to indicate that these specific packages are any better than many others. Due to size and single color limitations, only very simple examples can be presented. Vendor brochures, demonstrations, and sample plots and reports using real projects are the only way to gain familiarity with these products and their outputs. Rosenau[4] provides a few representative examples from two widely used packages. Levine[5] shows a number of examples, but the current package capabilities go far beyond his illustrations. Lowery's book[6] has many illustrations of the outputs from one widely used package, as does Halcomb's[7] for another.

These many changes, coupled with the continual, dramatic increase in the power of microcomputers while simultaneously reducing their cost significantly, have spurred a remarkable resurgence of interest in, and practical application of, the principles of project planning, scheduling, monitoring, and control that have been well known for many years. The power and ease of use of the microcomputer systems now available surpass the fondest dreams of the practitioners of the 1960s who struggled to make the mainframe systems of those years do the job. Users who gave up on that struggle and reverted to manually prepared plans and schedules have now returned to capitalize on the significant benefits of computer-supported PMIS.

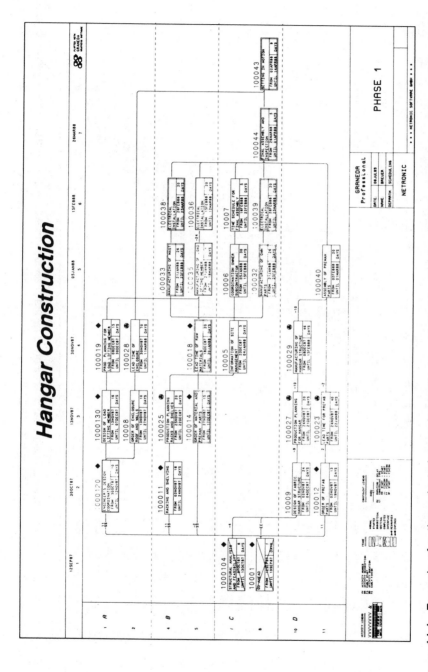

Figure 14.4 Example of a time-oriented precedence diagram plotted by the Graneda Professional software package. (Source: Netronic Project Management Graphics, Inc. Used by permission.)

307

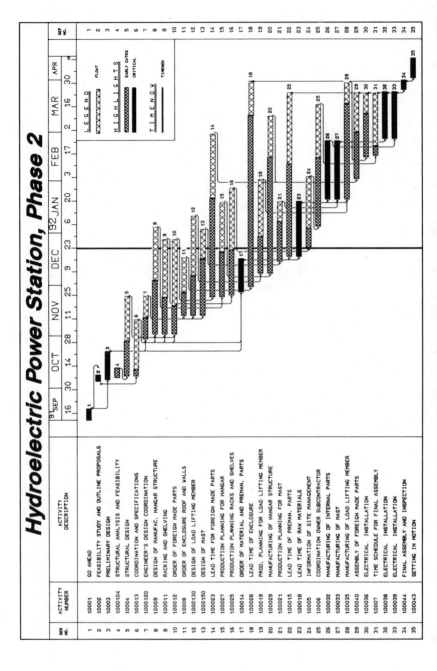

Figure 14.5   Example of a fixed time scale logic bar chart plotted by the Graneda Professional software package. (Source: Netronic Project Management Graphics, Inc. Used by permission.)

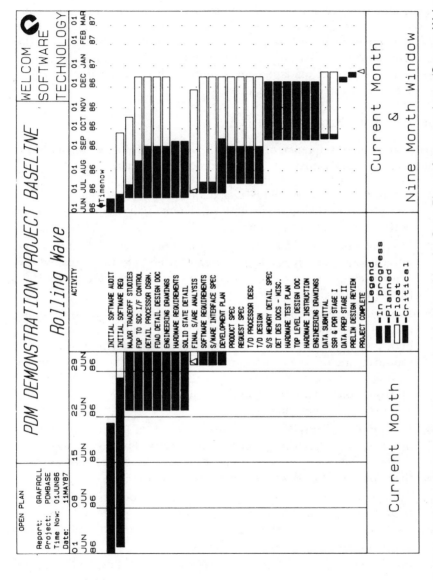

Figure 14.6 Example of a rolling wave bar chart schedule plotted by the Open Plan software package. (Source: Welcom Software Technology. Used by permission.)

309

14.3 SELECTION OF PROJECT MANAGEMENT SOFTWARE PACKAGES

The project management software market is large, rapidly growing, and highly competitive, as indicated in Figure 14.3. The market-size information shown in that figure includes only the North American market; there are also major markets in Europe, Asia, and the Far East. Selection of the best available software package from among the many competing products to support the project management activities within a specific organization is a complex and demanding task.

The Selection Process

The recommended sequence of steps to be followed for selecting a system of any size for mainframes, minis, or microcomputers is:

1. *Define the user needs.* This is a major task that frequently is not carried out in the depth or with sufficient analysis and thought for both current and future needs.

2. *List and rank the system specifications.* These should be separated into those absolutely required, those desired, and those wished for. They include the computer platform to be used and the system functions and characteristics.

3. *Issue request for proposal.* Depending on the situation, this simply may be a telephone call to selected software vendors, or it may be a highly formalized request. In either case, the defined user needs should be made available to prospective suppliers. A preliminary culling of the many offerings is usually required, on the basis of hardware platforms, price, and key functions, to avoid having 50 or 100 proposals to evaluate.

4. *Evaluate proposals received.* Usually some sort of scoring method will be needed to rank the competing proposals. The costs (hardware, software, labor—for installation, training, and usage) must be identified and compared with the expected benefits.

5. *Compare and rank the finalists.* For a small number of final contenders, conduct demonstrations and tests, preferably on the buyer's premises. Interviews should be conducted with current users of the candidate systems.

6. *Present findings and obtain needed approvals.* Purchase of one of the low-end packages for a few hundred dollars may not require very high level approval. However, implementation of a particular package will set a precedent from a systems viewpoint that may have unforeseen ramifications, and once a particular package is in use, it

is often very difficult to get people to change, even if a new package has many advantages. So it is wise to obtain the approval of various people in the organization, including other potential users, senior managers, and information system specialists and managers, even when one has the authority to approve the expenditure amount.

7. *Acquire and install the package.* For microcomputers, this is usually a simple, straightforward procedure, but for the larger systems, it may be rather complex. Getting the system running on your computer is only a first step. Structuring the coding schemes to be used in the system, for example, for the project breakdown structure, organization breakdown structure, and resource skills, to name just a few, is a major part of the implementation process.

8. *Train all affected and implement the system on specific projects.* Two levels of training are usually required: Specialized training of support specialists in the details and intricacies of the system operation, trouble shooting, modifications, and advanced usage; and training of the average users in entering data and using the output results for their day-to-day project management purposes. Frequently, organizations fail to get the most out of their new systems because inadequate training is provided at one or both of these levels. Even with the simplest, low-end microcomputer package it is desirable to have one person as the resident expert in the system.

The typical elapsed time for this selection and installation process is from three to six months. It can be accomplished in a much shorter time, especially in cases where there is only one user, or a few small projects, or when only the basic scheduling functions are required.

Factors to Be Considered by Purchasers of Project Management Software Packages

There are many factors to consider when selecting a project management software package. As indicated in Figure 14.7, a number of important factors are not directly related to the actual functions or operation of the package itself. Users have learned through bitter experience that there is a rather high mortality rate among software vendors, and the sudden disappearance of a vendor after considerable effort has been expended to implement and train your people in the use of their package is disconcerting and potentially expensive. This means that user support is not available, and that the package will not be improved and updated to take advantage of the rapid progress in computer technology.

Each of the product factors listed in Figure 14.7 must be further detailed in order to objectively compare several competing packages. As examples, two of the factors listed in Figure 14.7, Resource Data and Cost

1. *Vendor Reliability and Reputation*
   Length of time in business
   Reputation
   Size of organization
   Focused on project management
   Number of products
   Financial strength/reliability:
   Geographic areas providing sales and support
   International areas supported
   Client or reference list
   Training provided
   Consulting provided

2. *Product*
   General product information
   Synergy with other products
   Project definition data
   Work breakdown structure
   Scheduling features
   Activity definition data
   Resource data
   Reporting (C/SCSC, other)
   User interface (menu, graphics, mouse)
   Import/export data to other systems
   Calendars: number, use, date formats
   Outline format
   Data base architecture
   Monitoring data
   Cost data
   Graphic outputs

3. *Documentation*
   On-line and multiple level help commands
   Additional training materials
   Error correction procedures clearly documented
   Tutorial
   On-line tutorial
   Quick reference card

4. *User Support*
   User support dedicated telephone
   Toll-free (800) telephone line
   Workshops for special topics
   Notification to users of upgrades
   Newsletters
   24-hour hot line
   Training classes
   Consulting to users
   User groups
   Electronic mail

5. *Contract Issues*
   Package functions
   Site licensing
   Modifications to package
   Education and training
   Upgrade & enhancement policies
   Vendor liability
   Vendor bankruptcy (escrow)
   Licensee bankruptcy
   Confidential data
   Package efficiency
   Delivery dates
   Installation and acceptance
   Support commitment
   Warranty, maintenance agreements
   Price and terms
   Licensee defaults payment
   Proprietary interest

Figure 14.7   Typical factors considered by purchasers of project management software packages. (Source: Ellodee A. Cloninger, *Project Management Software Buyer's Guidelines* Cupertino, CA: Pmnet, 1988. Used with permission.)

Data, have been expanded in Figure 14.8. It is only at this level of detail that a purchaser can determine if the product will meet the specified user requirements.

Sources of Assistance in Package Selection

It is virtually impossible for the average user of these systems to keep abreast of the continual improvement of the many available and new software packages in this field. A number of computer-oriented magazines publish the results of their surveys and ratings of these packages, and several software rating companies offer specialized reports of their proprietary tests. The journals, proceedings, and other publications of the

---

1. *Resource Data*

    Number of resources per project
    Resources applied across activity:
    —uniformly
    —unevenly
    System notification of resource
    constraining the schedule
    Resource interrupts permitted
    Interactive graphic resource profiles
    Resource reports across multi-projects
    Change resource units during project
    Multiple billing rates per resource
    Absolute resource limits allowed

    Number of resources per activity
    Resource constrained schedules
    Resource levelling:
    —automatic
    —across multiple projects
    Priorities assigned to resources:
    —early or late dates
    —float/slack values
    —other
    Resource unit change reflected in
    schedule change

2. *Cost data*

    Types of cost data supported:
    —actual costs
    —committed costs
    —original estimates
    —revised est/how many
    —target costs
    Resource costing by:
    —total project
    —activity
    —time period
    —-prorated over time
    —work breakdown structure
    User override system calculated
    escalations

    Cost fields supported:
    —fixed
    —variable
    —user defined
    Cost summaries supported:
    —activity
    —resource
    —time unit
    —project/work breakdown
    structure
    —across multi-projects
    User defined multiple cost
    escalation rates
    Cost constraint schedules

Figure 14.8   Examples of typical detailed factors to consider when selecting a project management software package. (Source: Ellodee A. Cloninger, *Project Management Software Buyer's Guidelines* Cupertino, CA: Pmnet, 1988. Used with permission.)

several professional associations concerned with project management contain useful, up-to-date information on this fast-moving field. Examples of these publications include:

The Project Management Institute, P.O. Box 43, Drexel Hill, PA 19026-3190 USA:

The Project Management Journal (ISSN 8756-9728) (4 issues per year)

pmNETwork (ISSN 1040-8754) (8 issues per year)

Proceedings of the Annual PMI Seminar Symposium (1970 to present)

The PMI Project Management Body of Knowledge

Books, reports, and monographs.

INTERNET International Project Management Association, International Secretariat, Todistrasse 47, Postbox 656, CH-8027 Zurich, Switzerland:

Proceedings of the INTERNET World Congresses, Symposia, and International Expert Seminars

INTERNET Handbooks and Reports

INTERNET is an affiliation of 28 national associations (mostly European) devoted to project management, most of which publish journals and reports in their national languages.

The Association of Project Managers (APM), Secretariat, 85 Oxford Road, High Wycombe, Buckinghamshire, HP11 2DX, UK:

Project, the monthly bulletin of APM.

The International Journal of Project Management (ISSN 0263-7863)

Published by Butterworth-Heinemann Ltd., PO Box 63, Westbury House, Bury Street, Guildford, Surrey GU2 5BH, UK, for APM and INTERNET, 4 issues per year.

Performance Measurement Association, 101 Whiting St., Suite 313, Alexandria, VA 22304 USA:

Annual Proceedings

Newsletter (quarterly).

Focus is on U.S. DOD/NASA C/SCSC requirements.

American Association of Cost Engineers (AACE) and AACE-Canada, PO Box 1557, Morgantown, WV 26507-1557, USA:

Cost Engineering (ISSN 0274-9696) monthly

AACE Transactions, proceedings of the annual meetings

Cost Engineers' Notebook

AACE Recommended Practices and Standards

A number of consultants specializing in project management are available to assist in the selection process, together with the software vendors themselves. These consultants can be located through the professional associations previously listed, and through the Institute of Management Consultants (New York) and its affiliated management consulting institutes in other countries. Yahdav[8] provides one of the most comprehensive evaluations of the available microcomputer project management packages, including (1) a detailed listing of all vendors and systems, (2) a "shopper's guide" with advice on how to carry out the selection process, (3) his "PM Solutions Strategy," (4) the results of his extensive comparative analysis, (5) detailed analytical reports on the 18 top-ranked packages, and (6) a useful "Guide to Terminology." Yahdav's approach involves three steps: (1) testing each package against a comprehensive set of project management requirements made up of 350 parameters involving 1,200 criteria, (2) comparing and ranking the packages through evaluation of the test results into first, second and third choice categories, and (3) analyzing selected systems individually as solutions to typical project management situations. In early 1991, Yahdav listed six microcomputer packages in his first-choice list (basic, with dot-matrix graphics), and seven in his first-choice list of packages that support plotter graphics, ranging in price from $395 to $4200 on a single copy basis. (Volume discounts are significant, and packages for use on local area networks (LANs) are more expensive.) An additional 21 packages were listed as second-choice, ranging in price from $295 to $3500. New releases of all of the competing packages with additional features and improvements are issued so frequently that the rankings are constantly changing. Yahdev summarized his comparative test results using 17 aspects.

Overall Project Management with Graphics Power

1. Scheduling time frame.
2. Project modeling.
3. Scheduling management.
4. Resource management.
5. Cost management.
6. Government reporting requirements.
7. Advanced analysis.
8. System interface.
9. Control management.
10. Plotter-graphics program.
11. Top management information.
12. Scheduler tools.
13. Report writer (flexibility).

User Friendliness

14. Setup and learning.
15. Ease of use.
16. Ease of user interface.
17. Documentation completeness.

In summary, selection of computer software packages to support the project management discipline within a given organization requires a systematic and thorough approach to assure that the best possible choices are made. The amount of initial investment in hardware and software will be a small fraction of the total cost in managers' and planners' time expended in using the system over a period of years, and a few hundred dollars saved in the initial investment can cost many thousands of dollars in lost time, not to mention the impacts on the affected projects, if the wrong decision is made.

14.4   IMPORTANT FACTORS IN PMIS USAGE

Systems Do Not Manage Projects

Any information system, including the best project management information system, will not manage a project. The people involved in any project (general manager, project sponsor, project manager, functional project leaders, and other team members) do that. A good PMIS is a valuable tool that can greatly assist these people in managing their project. The project management triad—the integrative roles, project teamwork, and integrative and predictive planning and control systems—must be well-balanced and mutually supportive to be effective. People must read, absorb, understand, and take action on the information that the system presents to them. These points, which may seem obvious, are restated here to help avoid a common trap: Many managers believe that all they need to introduce or improve project management in their organization is the latest and greatest project management computer software package. They are disappointed and angry when six months after the package has been purchased and installed their projects are still out of control.

Improvements in an organization's project management capabilities must move forward on all three fronts: the integrative roles, project teamwork, and integrated planning and control. As discussed in Chapter 10, documenting and improving the organization's overall project management process is perhaps the most fertile field for improvement efforts in most organizations today. Automating a disjointed, inefficient, archaic process with the best project management package will produce only marginal benefits.

Project Planning Templates and Libraries

Many aspects of project action and resource planning lend themselves to development of templates or standardized modules that can be stored in hard copy or electronic libraries for future use. Although each project, by definition, is unique and has not been carried out precisely in the same manner before, many if not most of the tasks that make up the project are repetitious. The results of each task may be slightly different, and the duration and resources required may change somewhat, but the basic steps—the process—followed in each task are often identical with a previous task.

Systematic project definition using the project/work breakdown structure approach enables recognition and definition of many standardized breakdown modules. Key milestone and interface events are frequently identical from one project to another, even though the time between such events may vary considerably. At the task/work package level, the action network plans are at least 80 percent identical in the logical flow of work for similar tasks, even though the duration of some activities in the network may vary from project to project. Patterns of resource usage at the task level often are remarkably similar from one project to another.

Rather than ask each new project manager and project team to start with a clean sheet of paper (or blank computer screen) on each new project, it is much more efficient to provide the team with a library of planning templates which they can select, adapt, and link to form the plan for their new project.

In years past, network planning templates had to be printed in hard copy, perhaps using transparent adhesive film, which could then be pasted together and marked up manually to form an overall plan. Today, these templates are stored electronically in computer files. They can be called up on the computer screen and projected on a large screen in the team planning theater described in Chapter 11, modified by the team members in joint session, and linked to form an integrated plan. The team members can be given hard copies to review in detail, or they can return to their individual offices and call up the project planning file on their desktop computers to further refine the plan and proceed to the next lower level of detail in the schedule hierarchy. Alternatively, they can carry the files with them on a disc, insert this into their own computer, and proceed with their planning work.

The best experience and most efficient work flow, based on many prior projects and inputs from seasoned veterans, can thus be handed to the less experienced project managers and team members for use in creating their project plans. There is always a danger in this approach of having too much standardization, and depending too much on established routines, which would stifle creativity and improvement. The project teams must use these templates as guides only, and constantly look for ways to better them. When such improvements are made, they can then be made a part of the planning template library, and all others in the organization will then immediately have the benefit of the improvements.

These electronic files can easily be transmitted anywhere in the world using electronic mail (E-mail) systems, or by direct computer-to-computer file transfer protocols. While there is still a place for telephone facsimile transmission of some graphic displays, E-mail allows a faster transmission of larger amounts of information in a shorter time, with more legibility, assuming that the receiver has the appropriate computer plotting devices to produce the graphic results.

Audiographic teleconferencing, using standard microcomputer hardware and a high quality sound system to share audio and visuals with many people simultaneously and interactively, promotes teamwork no matter how many miles separate the project team members. With this technique, voice and computer screen images, including pictures, text, data, and high-resolution graphics are transmitted live to participants in a team planning session, for example, at multiple locations over ordinary phone lines. This approach is therefore less expensive and more practical than video conferencing, which requires videocameras and video transmission circuits. With audiographic teleconferencing, using an electronic pen and tablet, team members can create or annotate visual materials on the computer screen, and several people can speak and use their tablets at the same time, creating a high degree of interaction.

Future Possibilities

Although it is risky to predict with any certainty what the future developments will be in a field that is moving as rapidly as information technology, through extrapolation of trends it is fairly safe to say that:

- Integrated project management information systems will be increasingly recognized as fundamental tools required by any organization responsible for project-like efforts.
- PMIS will continue to gain in power, scope, and automated integration with the other related information systems in the organization.
- Multiproject operations planning and control systems of the type described in Chapter 8 will become increasingly more common.
- Information technology advances, including audiographic teleconferencing and multimedia concepts, combining text, data, graphics, pictures, and full-motion, color video images with touch-screen access and E-mail, will:

    Enable managers from the chairman down to access and interrogate their multiproject PMIS files,

Encourage and enhance project team planning and integrated project evaluation no matter how widespread geographically the team members may be,

Be used for training project managers and project teams using pertinent projects to simulate real-life situations in the industry and organization involved,

Enable development of useful expert systems for project planning, scheduling, monitoring, and control, capturing the total experience of an organization and industry.

The demand for more powerful, more integrated and more easily used project management information systems will continue to grow for the foreseeable future, and these systems and effective use will be major determinants of who will come out ahead in the competitive global marketplace.

15

◀ ▶

Evaluating and Directing the Project

Once a project is set in motion the project manager must continuously monitor all facets of project activity in conjunction with the contributing functional project leaders. The project manager must evaluate the project in its totality and initiate, with and through the functional project leaders, appropriate directive actions to recover from or prevent undesirable results. This chapter deals with this ongoing process.

15.1 INTEGRATED PROJECT EVALUATION: NEED AND OBJECTIVES

Each area of cost, schedule, and technical performance can be monitored and evaluated separately by one means or another. However, complex interrelationships exist between these three areas. The need is to evaluate the total project in all three areas simultaneously, on an integrated basis. Technical problems cause delays; delays increase cost; budget overruns may adversely affect quality, and so on. Corrective actions in one area of the project may cause unforeseen problems in another area.

The *objectives* of integrated project evaluation are the following:

- To *provide visibility*, as clearly as possible, of the interrelationships between cost, schedule, and technical performance across the entire project.

- To *identify problems before they occur* to the extent possible, so that they can be avoided or their effects minimized.
- To *identify opportunities* quickly for schedule acceleration, cost reduction, or technical advance, and to exploit them before the opportunity is lost.

15.2 METHODS AND PRACTICES OF PROJECT EVALUATION

Project evaluation is a continuing process throughout the life cycle of the project. It should not be an occasional panic exercise triggered by sudden awareness of a major problem.

The Project Evaluation Process

Project evaluation is a three-step process, repeated at periodic intervals.

1. *Determine status* on a total project basis, of

    Actual work accomplished

    Current and anticipated technical results

    Resources expended (time, labor, money).

2. *Compare status to plan,* for

    Schedules

    Budgeted and currently estimated costs

    Technical specifications to be met, both at the time of comparison and at completion of the project.

3. *Identify variances* between current or future cost, schedule or technical performance, and related plans.

At this point, the evaluation process gives way to project direction, which involves determining and authorizing appropriate actions to eliminate the variances between performance and plan. Further evaluation may be necessary, however, of proposed or alternative actions prior to their implementation.

Basic Methods and Practices

In the following sections six basic methods and practices for project evaluation are discussed.

- First-hand observation.
- Interpretation of verbal and written reports.

- Graphic display of information.
- Project evaluation review meetings.
- Project performance reviews.
- Project control center.

First-Hand Observation

It is vital that the project manager have personal contact with as many contributing functional team leaders and specialists as is possible. Direct observation and contact with the functional team leaders and other project contributors is invaluable to determine physical progress, whether this is related to a design drawing or field installation, and to identify potential problems in those particular fields where the project manager is experienced.

There are definite limitations on the effectiveness of first hand observation, however, especially on large major projects. Geographic dispersal and sheer size and complexity often limit the amount of first hand observation a project manager can make on a periodic basis. Even under ideal conditions, additional information and evaluation are required.

Interpretation of Verbal and Written Reports

Verbal reports have the great advantage of being very current, but they also have a great disadvantage in that they are easily misunderstood, distorted, and forgotten by both parties. Accountability is almost totally absent.

Written reports and documents are valuable, and some are indispensable for project evaluation. However, these are frequently less effective than they could be because

- Many reports are poorly designed, too detailed, and difficult to understand.
- With numerical tabular reports, it is difficult to identify significant points of change and trends.
- Written technical progress reports may describe in great detail what was done during a given period, but seldom do they present significant information on current or future problem areas and progress toward meeting the ultimate product specifications.

Some steps that can improve the effectiveness of report interpretation include:

- Proper summarization for the purpose at hand (using the project breakdown structure, for example, to summarize expenditure information, not simply a standard chart of accounts).

- Comparisons and ratios.
- Comparison with previous reports to show trends.
- Selective reporting of only the information of interest to the recipient.
- Conversion to graphic display.

Graphic Displays

Graphic display of the key elements of information concerning a project improves the manager's ability to evaluate the project on an integrated basis. Properly designed graphic displays are effective because

- Large amounts of complex information can be presented in pictorial, easily understood form.
- Changes in rates of progress or expenditure are easily identified.
- Different kinds of information (schedule, cost, and technical) can be presented simultaneously for integrated analysis.
- Original plans, past performance, and future predictions are easily shown and compared.
- Open display, if possible, maintains an awareness of the project and its progress on the part of all people concerned.

Graphic display has the following disadvantages:

- Extra effort is required to design, prepare, and maintain various charts and graphs.
- Display of information at too gross a level may mislead management by hiding problems.

Back-up, detailed reports are necessary to prepare and maintain graphic displays and to carry out in-depth analysis of apparent problems indicated by such displays. Several of the figures in Chapters 10 and 12 are useful for display purposes. In addition, examples of typical display reports are given in Figures 15.1, 15.2, 15.3, and 15.4. A number of variations on these basic charts are possible.

As discussed in Chapter 14, one of the important advances in project management information systems over the past decade has been in the graphic presentation of project planning, scheduling, cost, and resource management information. Today the cost of preparing and using graphic displays as described earlier and illustrated throughout this book is too small to be a serious deterrent to their use. This graphic display capability is probably the single most important factor in gaining higher level manager understanding and use of integrated planning and control systems for project management. It is equally important in

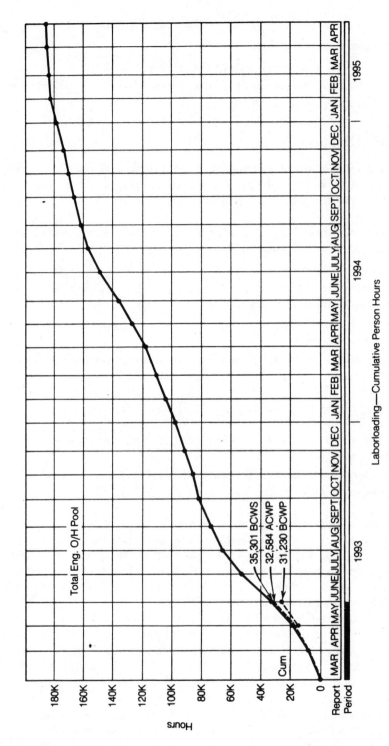

Laborloading—Cumulative Person Hours

Figure 15.1   Labor loading chart.

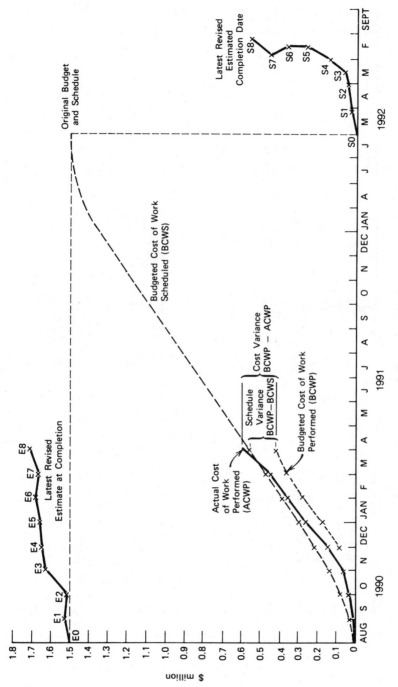

Figure 15.2 Cost and schedule performance chart.

325

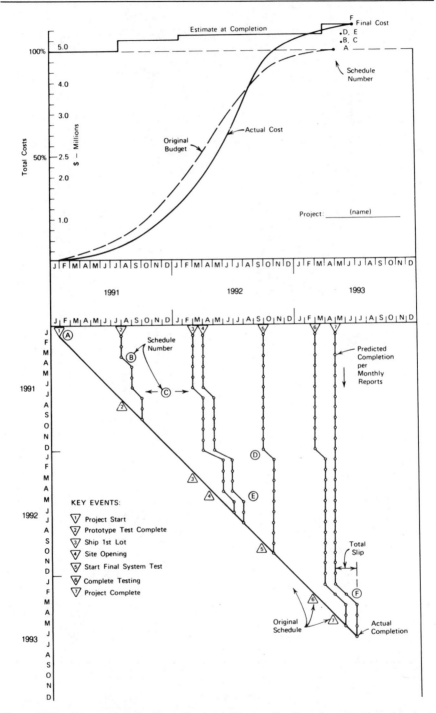

Figure 15.3   Cum-costs milestone slip chart. Shows situation at completion of project.

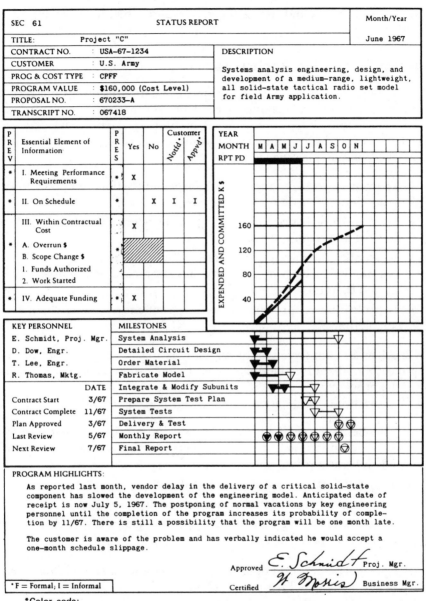

Figure 15.4   Example of a program status report. (Source: "Multi-Project Control," Robert A. Howell, *Harvard Business Review,* March–April 1968. Used by permission.)

gaining acceptance and use of these disciplines by the average project team member.

Project Evaluation Review Meetings

The holding of regular, periodic project evaluation meetings is a recommended practice for all major projects. Here key managers are confronted with the current, integrated analysis of progress against the plan and predictions for the remainder of the project. The primary purpose of project evaluation meetings is

- To identify problems requiring management action, or opportunities to accelerate schedule or reduce costs.
- To obtain a real agreement among key managers that these problems do exist and require action.
- To identify the manager responsible for action to resolve each problem or exploit each opportunity, and to record action assignments related thereto.
- To follow up on previous action assignments to assure their early completion.

Problems are not solved in the review meetings. Separate problem solving meetings should be held, attended by only those concerned directly with the problem.

Effective project evaluation review meetings require that

- They be held weekly for most projects, or at other regular intervals on a preestablished schedule, so key participants will plan their schedules accordingly.
- Designated participants personally attend.
- They be brief and follow a prepared agenda.
- Written action assignments, with specific names of people and due dates, be prepared and distributed within 24 hours.

Persons participating should have commitment making authority and include key members of the project team:

Project manager (usually chairman)
Key functional project leaders or managers
Project engineer
Project contract administrator
Project controller

Project accountant

Key staff specialists

This regular, periodic confrontation of all project team members, with personal statements by each on his area of responsibility, can generate a strong team spirit and personal commitment by every team member. If the project manager tries to dominate the meeting, the team members are likely to close off communication and let the manager try to discover what the problems are without their help.

Project Performance Reviews

Another form of project review, similar to a management audit, is conducted by a team of experienced project management professionals who are outside the project team. Such reviews are conducted at critical points in the project life cycle and, if conducted properly, can be beneficial to assuring the ultimate success of the project. Ono and Archibald[1] describe the experience in AT&T's Business Communications Systems (BCS)* group with such reviews, whose objectives are:

1. To assist the project team in bringing in the project on time, within budget, and within the defined scope.
2. To determine adherence to AT&T's Project Management Guidelines, with emphasis on project management deliverables and processes.
3. To provide a record for dissemination of the lessons learned.
4. To provide data for validating the effectiveness of AT&T's project management education and training programs.

The results of the evaluation are intended to enhance AT&T BCS Project Management execution and:

- Improve the immediate performance of evaluated projects where required.
- Validate and refine BCS project management standards and guidelines for use with future projects.
- Provide an input for individual performance appraisal.
- Highlight strengths and identify areas for improvement in regard to project activities and the performance of the functional and/or project organizations.

*Refer to Section 11.3 for a description of the types of projects involved in these reviews.

- Identify innovative techniques that enhance project success and client satisfaction, so they can quickly be transferred to other projects.
- Produce additional understanding of the timing and level of project management resources required to produce the project management deliverables required on each of the AT&T projects.

The AT&T performance review process consists of the following steps:

1. Identify candidate projects and select the next project to be reviewed.
2. Form the Project Review Team and schedule the review.
3. Conduct the project review.
4. Provide immediate on-site assistance as required.
5. Provide verbal feedback and prepare the written evaluation review report.
6. Prepare a success assurance plan, if required.
7. Monitor implementation of the success assurance plan.

The Project Review Team consists of the National Director of Project Management, an area Project Director, and a Project Manager, both from an area outside that of the project under review. The reviews normally require three days of intensive effort, and encompass examination of the project planning documents and interviews with the project manager and other key project participants. Guidelines and lists of interview questions are provided in the BCS Project Management Guidelines.

AT&T concluded that the project evaluation review process is of ongoing value as it (a) identifies projects in trouble with enough time left to correct the problems before a disaster strikes, (b) provides a mapping for the project team to get out of trouble, (c) enhances the adherence to the project management guidelines even for projects not being reviewed, (d) provides the "best training for project managers" (as one of the project managers under review has stated), (e) provides a formalized, regularly scheduled method for continued updating of the project management process and guidelines, and (f) serves to strengthen the reliance on and cross pollination of the informal project management network across the country. Consequently, this review process provides a formal and an informal communications path for lessons learned within the organization.

Project Control Center

The practice of establishing a special room or center for the display and evaluation of project information began in the 1960s, and the effectiveness of this practice is indicated by its continuing adoption by companies in many industries. Control centers are set up for a single large project or program, many projects, or, in some cases, the total company operation.

A project (or management) control center serves the following purposes:

- Provides a single location for concentrated display of relevant information about the project (or projects).
- Serves as a physical representation of the project or projects, reminding all concerned of its existence, status, state of health, and importance.
- Serves as the meeting place for evaluation review meetings and other conferences necessary to the project, providing ready reference to facts pertinent to aspects of the project under discussion.
- Provides through simultaneous displays of graphic, tabular, narrative, and pictorial information a most effective means for the project manager and others to integrate this information mentally, apply judgment and intuition, and identify current or potential problems requiring action.

Suggested facilities to be provided in the project control room include:

- Wall display panels (may be sliding or rack-mounted) to hold charts and graphs.
- Tackboard for temporary display.
- Blackboard or white board (magnetic can be useful for developing network plans or displaying charts).
- Computer terminals (micro computer and/or mini/mainframe terminal).
- Screen and projector (computer driven, transparency overhead, other).
- Television facilities as appropriate.
- Conference table with seating adequate for the size groups expected.
- Storage facilities.
- Desk and drafting table and facilities for recording proceedings and preparing charts and transparencies (may be in separate room).
- Telephone facilities to support telephone and audiographics teleconferences.

Size of the project and other considerations will dictate how elaborate the facility should be. It is important that such centers be actually used for team meetings, or they will simply become show-places to impress customers or visitors.

15.3   DESIGN REVIEWS AND PRODUCT PLANNING REVIEWS

Design Reviews

At key milestone events, design reviews are conducted to assure that the technical objectives have been or will be met. To be most effective these reviews are performed by persons outside the project, and even outside the division or company. This assures that an independent audit of the adequacy of the design is obtained. Such reviews are planned and scheduled by the project manager and conducted by an independent chairman not associated with the project.

Engineering inspections of prototypes, mock-ups (full-scale models), pilot plants, or other types of models of the end result are also a valuable means to determine the adequacy of the technical or engineering effort prior to full commitment to the following phases of the project (construction, manufacture, field erection, etc.).

Product Planning Reviews

For major product development projects, the project will have been justified by a product plan, with emphasis on the economic factors, such as product price, market, sales volume, competition. At key milestone events in the life cycle of the development project, the entire product plan must be reviewed. Such reviews will reflect the progress to date on the project, currently predicted product and project cost, changes in market and competition, and so on. Failure to conduct such reviews after a major project has been authorized has resulted in development of a number of excellent products for which the market has disappeared, either due to product cost increases, delays in project completion, or competition and market changes.

15.4   PROJECT DIRECTION

Integrated project evaluation, as described earlier, is needed so that proper direction of the project can be exercised. Project direction includes all forms of communicating in order to

- Change the course of actions that have caused or will cause problems or undesirable cost, schedule, or technical results as follows:

     Replan to recover to schedule and/or budget.

     Initiate effort to correct technical difficulties.

     Reallocate budgets (money, manpower) as required.

- Alleviate the unavoidable undesirable results by the following:

   Document customer delays.

   Invoke force majeure where possible.

   Negotiate new schedules, prices, or budgets.

   Revise technical objectives.

   Negotiate changes in scope of work.

The project manager is similar to the pilot at the wheel of a ship or at the controls of an aircraft. He or she continuously monitors the progress of the project through the evaluation system, watches for indications of present or future difficulty, and communicates to the appropriate functional specialists any need to change plans, schedules, budgets, and performance to reach the project objectives. The "pilot" must also monitor these change signals to be sure they have been received, understood, acted upon, and that they do in fact produce the desired result.

The primary means of giving internal project directing instructions are:

- Action assignments resulting from project review meetings.
- Project directives, special memoranda or E-mail messages.
- Task work orders, subcontracts, PAR, research and development case, product plan release and revision documents, or contracts and contract amendments.

Verbal directives affecting cost, schedule, technical performance, or scope of work must be documented by one of these means.

Direction through Action Assignments

The primary document produced by the project evaluation and review meetings is the *Action Assignment List.* Each specific identified problem or issue related to cost, schedule, or technical performance is addressed by an action assignment. The project manager defines and records each action assignment, specifying:

- *What* the problem is.
- *How* it is to be solved, in general terms (if possible).
- *Who* has the lead responsibility for the action, and who will contribute.
- *When* the action is scheduled to be completed.

The Action Assignment List is signed by the project manager and distributed within 24 hours after each project review meeting to all key project team members and functional managers. E-mail is very effective here. Accountability is pinpointed, and good coordination is assured since everyone is informed of all action assignments. Any objections to the assignments can be voiced during the review meetings, and acceptance of the assignment in full view of the project team reinforces each person's commitment to complete it as scheduled. Such commitment to peers is much stronger than a one-on-one commitment to the project manager alone.

Continual follow-up must be exerted by the project manager on all outstanding action assignments, to assure their timely completion.

Project Directives

In some situations, it may be desirable to establish a procedure for issuing project directives as a means for documenting and disseminating policies and decisions. Normally these would be signed by the project manager, or for major actions by the project sponsor. Memoranda, letters, or E-mail messages can also be used effectively for giving directive instructions to project team members and contributors.

Direction through Task Work Orders, Contracts, and Similar Documents

Control of funds is, in most cases, the most effective means of directing and revising work scope, schedule, and cost. Approval for release or revision of task work orders, subcontracts, contract amendments, PAR's, research and development cases, product plans and similar documents provides a fundamental source of direction giving authority.

Changes to work scope, schedule, cost, or technical objectives resulting from action assignments or project directives, should be documented formally. Procedures are described in Chapter 12 to accomplish this.

15.5 REPORTING TO MANAGEMENT AND THE CUSTOMER

Company management must be informed continually of the overall health and status of the project. Problems that may jeopardize the schedule, profit, or budget must not be withheld from management, and possibly should be communicated with care to the customer. The status reports described here are intended to reveal certain of these problems to make higher level management aware of the actual or potential difficulties, either to assist in their resolution or take appropriate actions to minimize the adverse impact at the multiproject or corporate levels.

When reporting a problem to the project sponsor or to higher management, the project manager should state:

- The problem.
- The cause.
- The expected impact on schedule, budget, profit, or other pertinent areas.
- The action taken or recommended and the results expected of that action.
- What top management can do to help.

Reporting to the customer must be carried out in accordance with the terms of the contract. Careful coordination is required between the project manager, the contract administrator, and the commercial or marketing manager to assure that the content of reports to customers and the means by which they are handled serve the best interests of the company.

Reports to Management

For major commercial and governmental projects, as defined in Chapter 2, three levels of management reporting are defined

	Applicable to	Frequency
Major project register	All major projects	Quarterly or monthly
Monthly progress report	Designated major projects	Monthly
Critical project review	Critical projects	As scheduled

Major Project Register

As mentioned in Chapter 1, this is a summary listing of all major projects meeting the criteria established by appropriate corporate policies (similar to the criteria presented in Chapter 2). Figure 15.5 shows the information listed in a typical major project register. Detailed implementing instructions are needed to assure all reporting organizations comply properly with the policy.

The major project register performs the following valuable functions:

- Provides an up-to-date inventory of major projects, with related key data for each, for the information and evaluation of top management
- Identifies whether or not a project manager is in place on each major project

Company:
Product Line:

Serial No. (1)	Project Name and Description (2)	Project Manager (3)	Customer (4)	Value $000 (5)	Possible Exposure $000 (6)	Date of Contract Award (7)	Date of Contractual Completion (8)	Associated PARs (9) I.D. Number	Associated PARs (9) $000 Amount	Management Approval of Proposal Date (11)

1. Allocate a number, starting at 1 at the begining of each year; e.g., 93 (1) for the first project in 1993. This number is retained throughout the project life.

2. Project name and brief description of products and services comprising contract.

3. Give name and percent of time devoted to project.

4. Name (and location if export).

5. Gross charges to be billed through full life of project (in U.S. dollars).

6. Maximum penalty, damages or other exposure in event of default (U.S. dollars).

7. Date of contract award.

8. Contractual date of project completion.

9. Company number and amount of any capital project authorization requests (PAR) judged to be critical to the success of the project—e.g. new manufacturing facilities, test equipment, etc.

10. Company case number and amount of R and D cases, local and general, contributing significantly towards success of project.

11. Date of approval of proposal terms by responsible manager.

Figure 15.5 Typical program/project register.

- Assists in identification of the projects that have the greatest potential risk and exposure, so that management attention can be focused on these.

Monthly Progress Report

Cognizant executives at various levels may designate major projects requiring monthly progress reports, with appropriate distribution. The following is a suggested outline for such monthly reports:

1. *Summary status*—brief paragraph highlighting current status of the project.
2. *Red flag items*—previous and new red flag items, corrective actions taken, with prediction on resolution and further action required.
3. *Project staffing plan*—showing key or limited resources.
4. *Major achievements and future schedule*—describing actual accomplishments during current reporting period and significant changes in future schedule.
5. *Current and future problem areas*—stating major problems, actions required, and possible impact on the project.
6. *Project cost performance*—commenting on current project cost situation with reference to current cost performance reports.
7. *Exhibits*—(A) Summary Master Schedule (where required), (B) Detailed Project Schedule(s), (C) Project Cost Performance Report.

Many of the report formats described in Chapters 10, 12, and 15 are useful for inclusion in such monthly reports.

Critical Project Review

For specified critical projects, review meetings are held with appropriate executives to assure that the project is progressing as scheduled. These may be monthly, bimonthly, or quarterly, and they may be held on site or at division or higher headquarters levels.

Possible exhibits for projection and distribution in such reviews would include:

1. Project identification data.
2. Summary status, red flags, problem areas.
3. Summary master schedule.
4. Cumulative costs, milestone slip chart.

5. Cost performance report chart.
6. Action assignment summary.
7. Project manpower status.

Reports to the Customer

Good project management requires periodic (usually monthly) reporting to the customer.

These reports should be in writing but should be presented personally by the project manager accompanied by appropriate contracts (marketing or commercial) and legal representatives. Such monthly reports provide the opportunity to document delinquencies or requests for change on the customer's side that may affect the project schedule, cost, or scope. Other factors such as strikes, floods, supplier delays, shortages, and the like also can be documented in these monthly reports. Such documentation is vital at project completion to quick settlement of claims, in or out of court. The content and format of such reports depends upon the situation within each project and for each customer.

16

◀ ▶

Project Close-Out
or Extension

By definition, a project has a definite end point. However, terminating or closing out a project is more easily said than done. The project manager has the job of literally putting him or herself and the project team out of business. This is a very demanding assignment and requires considerable discipline to achieve a complete termination on the established date, especially if the manager's next assignment has not yet been identified.

Projects have a tendency, for these and other reasons, to stay alive long after their scheduled completion dates. There may only be a few people remaining on the project, but a long termination period can cause an otherwise successful project to move from financial success to failure. As long as an open work order remains, continuing cost charges are made to the project.

Some companies have found it necessary to replace the project manager near the end of the project with a person who is skilled at closing out projects, to assure that the schedule is met. During termination of a contractual project, the contract administrator plays a very important part.

This concluding chapter covers these major topics:

- Close-out plan and schedule.
- Close-out checklists.

- Responsibilities during close-out phase.
- Project extensions.
- Post-completion evaluation.

16.1 CLOSE-OUT PLAN AND SCHEDULE

As the final close-out phase approaches, the project manager establishes a specific plan and schedule for project termination. This plan covers at least these major items:

- *Contract.* Delivery and customer acceptance of products and/or service, and completion of all other contractual requirements.
- *Work authorization.* Close-out work orders and assure completion of all subcontracts.
- *Financial.* Collection from customer and closing of project books.
- *Personnel.* Reassignment or termination of people assigned to project office or project team.
- *Facilities.* Close office and other facilities occupied by the project office and team.
- *Records.* Deliver the project file and other records to the appropriate responsible manager.

16.2 CLOSE-OUT CHECKLISTS

Appendix B presents checklists to assist in preparation of the close-out plan and schedule. These are similar in use and format to the project start-up checklists in Appendix A, and are presented as a starting point for each organization to develop its own checklists, reflecting the needs of its specific projects.

16.3 RESPONSIBILITIES DURING CLOSE-OUT PHASE

The project manager and the project contract administrator have prime responsibilities during the close-out phase, as indicated in Chapter 9. These responsibilities are repeated here for emphasis.

Project Manager

- Insure that all required steps are taken to present to the customer adequately all project deliverable items for acceptance, and that project activities are closed out in an efficient and economical manner.

- Assure that the acceptance plan and schedule comply with the customer contractual requirements.
- Assist the Legal, Contract Administration and Marketing, or Commercial Departments in preparation of a closeout plan and required closeout data.
- Obtain and approve closeout plans from each involved functional department.
- Monitor closeout activities, including disposition of surplus materials.
- Notify Finance and functional departments of the completion of activities and of the project.
- Monitor payment from the customer until all collections have been made.

Project Contract Administrator

- At the point where all contractual obligations have been fulfilled, or where all but longer term warranties or spare parts deliveries are complete, assure that this fact is clearly and quickly communicated in writing to the customer.
- Assure that all formal documentation related to customer acceptance as required by the contract is properly executed.
- Expedite completion of all actions by the company and the customer needed to complete the contract and claim final payment.
- Initiate formal request for final payment.
- Where possible, obtain certification from the customer acknowledging completion of all contractual obligations and releasing the company from further obligations, except those under the terms of guaranty or warranty, if any.

16.4 PROJECT EXTENSIONS

In many projects, such as new products, system development, and commercial projects of various kinds, the project manager may have a very important responsibility to obtain legitimate, funded extensions to the project, in cooperation with the Marketing Department. These would be to develop a new feature or otherwise perform work beyond the scope of the original project. It must be clearly recognized that such extensions do change the original scope, and in fact represent new, probably smaller, projects springing out of the original one. Each extension must be planned, scheduled, and controlled—and closed-out efficiently when the new scope of work is complete. Any request by the customer for a change may become a source of new business or extension of the project. Strict monitoring and control of the contract scope, as discussed in Chapter 12, will protect the financial objectives of the project and provide opportunities for legitimate project extensions.

16.5 POST-COMPLETION EVALUATION OR AUDIT

An often neglected but important final step in effective project management is the post-completion evaluation. It is frequently said that we learn by experience. However, the first things we forget are the unpleasant, bad experiences. This is very good for our individual sanity, but not very good for improvements in managing future projects.

Unless a formal post-completion evaluation is made on each major project (or representative smaller ones), we will not learn by experience, but will repeat the same mistakes a number of times, to the detriment of our organization. Looking back on a project without a thorough post-completion evaluation, it is quite easy to come to a general conclusion that, "Well, it didn't really go too badly, after all."

Mistakes are made on every project. Some are large, some small. Post-completion evaluation is required to:

- Identify the mistakes.
- Determine their impact.
- Determine how they can be avoided on future projects.
- Initiate appropriate changes and improvements in project management and functional policies and procedures.

In Chapter 1, some symptoms and probable causes of poor performance on projects are described. These provide a starting point for identification of problems and improvement opportunities during the post-completion evaluation. Such an evaluation should be conducted a reasonable time after completion, allowing enough time to elapse to be able to judge the ultimate success of the project, but not so much time that records are lost and memories fade completely. Perhaps one to three months after completion is the proper interval. The evaluation should be conducted by a person outside the project, to ensure objectivity. However, the person must be familiar with the project and the organization, and must have access to key project team members.

The general procedure for a post-completion evaluation is

1. Determine the original and final objectives, in terms of performance (end product), cost, and schedule.
2. Determine whether these objectives were met.
3. Where things went right, determine what factors contributed to the success.
4. Where things went wrong, determine the basic causes.
5. Develop policy and procedure changes to eliminate causes of missed objectives or other problems.
6. Implement the changes.

Actually, this effort can be considered a management project itself, and therefore should be managed using the principles presented in this book.

APPENDIX A

◀ ▶

Project Start-Up Checklists

A rapid, efficient project start-up can prevent the loss of significant amounts of time which will be costly to recover in later stages, as illustrated previously in Chapter 2. The project start-up checklists presented in this appendix are intended to help get the project moving at full speed as quickly as possible.

Start-up checklists provide a number of benefits to the project. They:

- Clearly indicate the start-up functions and responsibilities, reducing ambiguity and uncertainty.
- Reduce oversight of important factors.
- Permit start-up progress to be monitored.
- Aid project team members with little or no previous experience.
- Inform project team members about the activities of other participants, to facilitate rapid, smooth start-up and good continuing coordination.

Figure A.1 contains 26 checklists, as indicated in its index. These are presented as a starting point for each organization to develop its own checklists reflecting the particular needs of its specific projects.

Index

A.   Project Office (PO) and Project Team (PT) Organization
B.   Instructions and Procedures
C.   Financial
D.   Project Definition
E.   Project Master Plan, Budget, and Schedule
F.   Task Plans, Budgets, and Schedules
G.   Work Authorization and Control
H.   Project Evaluation and Control
I.   Management and Customer Reporting
J.   Marketing and Contract Administration
K.   Government Furnished Equipment and Services (GFE)
L.   Extensions—New Business
M.   Project Records Control
N.   Purchasing and Subcontracting Liaison and Policies
O.   System Engineering and Planning
P.   Project Engineering
Q.   Design and Development
R.   Project Integration and Equipment Support
S.   Manufacturing Policies and Liaison
T.   Reliability—Product Assurance
U.   Engineering Documentation
V.   Construction and Emplacement/Installation
W.   Spare Parts
X.   Site Technical Operations
Y.   Site Administration Operations
Z.   Transportation & Logistics

Figure A.1a   Project start-up checklists index.

PROJECT TITLE _____   AWARD DATE _____

CONTRACT NO. _____   COST TYPE _____

CUSTOMER _____   PROJECT MGR. _____

THE PROJECT START-UP CHECKLISTS ARE DESIGNED FOR USE IN THE FOL-
LOWING MANNER:

COLUMN I—ITEM NO.—Each task listed is identified by a specific number and grouped
into categories. Categories are based on functions, not on organizations or equipment.

COLUMN II—TASK DESCRIPTION—Task descriptions are brief. Tasks that could apply
to more than one category are listed only in the most appropriate category.

COLUMN III—REQUIRED, YES OR NO—Check whether the item listed applies to the
project.

COLUMN IV—DATE REQUIRED—Insert the required date for accomplishment of the
task.

COLUMN V—ASSIGNED RESPONSIBILITY—Insert the name of the man responsible
to see that the task is accomplished on schedule. This may be a member of the Project Office
or an individual within a functional department.

COLUMN VI—PRIORITY (PR)—A priority system established by the Project Manager
may be used here, e.g., Priority 1 may be all tasks that must be accomplished by the end of the
first month of the contract, Priority 2 by the end of the third month, etc.

COLUMN VII—NOTES, REFERENCE—Refer in this column to any applicable procedure, a government specification that may apply to that task, etc.

Figure A.1b   Project start-up checklists instructions.

ITEM NO.	TASK DESCRIPTION	REQUIRED		REQUIRED DATE	ASSIGNED RESPONSIBILITY	PR.	NOTES REFERENCE
		YES	NO				
A.	*PROJECT OFFICE (PO) AND PROJECT TEAM (PT) ORGANIZATION*						
1.	Establish the Project Skill Requirements						
2.	Define the Objectives of the PO and PT Organization						
3.	Define Specific Tasks to be Performed						
4.	Clarify and Write the PO Charter						
5.	Establish & Schedule PO Resource Requirements: Manpower, Facilities, Material, Etc.						
6.	Staff the PO with Personnel						
7.	Define Charter to Each Member of PO and PT						
8.	Assign Individual Performance Responsibility						
9.	Prepare the PO and PT Organization Chart						
10.	Conduct the Management Kickoff Meeting						
11.	Clarify Relationships with Functional Org.						
12.	Physically Consolidate PO Personnel						
13.	Establish Personnel Hiring Requirements & Control						
14.	Make Use of Extra-Divisional Functions, e.g. Central Engineering						
15.	Prepare Project Responsibility Matrix						
B.	*INSTRUCTIONS AND PROCEDURES*						
1.	Establish Project Image, Letterhead, Motto, etc.						
2.	Establish a Project Instruction System Issue Instructions For:						

348

3. Project Office and Team Organization
4. Shop Orders & Responsibilities
5. Documentation Issuance & Control
6. Scopes of Work
7. Technical Specifications
8. Trip & Conference Reporting
9. Project Technical Control
10. Engineering Documentation
11. Reporting System—Internal & External
12. Security Policy
13. Patents Policy
14. Drafting Policy
15. Purchasing & Subcontract Policy
16. Priorities & Emergencies
17. Scope and Funding Authorization Policy
18. Equipment Identification
19. Inspection Procedures—Internal & External
20. Packing, Shipping, Logistics Policy
21. Spare Parts Policy
22. Reliability & Product Assurance Policy
23. Test Procedures & Acceptance
24. Project Meetings & Communication Media
25. Field & Site Support Operations
26. Set up Project File

C. FINANCIAL

1. Issue Master Contract Release
2. Establish Approved Contract, Case and/or Par Cost

349

Figure A.1c   Project start-up checklist.

ITEM NO.	TASK DESCRIPTION	REQUIRED		REQUIRED DATE	ASSIGNED RESPONSIBILITY	PR.	NOTES REFERENCE
		YES	NO				
3.	Develop Profit Trade-Offs on Incentive Contracts						
4.	Establish Project Direct Budget						
5.	Establish Project Indirect Budget						
6.	Issue Project Release Document						
7.	Establish Project Funding Plan—Customer & Internal						
8.	Obtain Marketing and Engineering Assistance Budgets						
9.	Establish Proposal Funding Policy						
10.	Establish Project Chart of Accounts						
11.	Set up Management Reserve Transaction Register						
D.	*PROJECT DEFINITION*						
1.	Establish Approved Contract, Case and/or PAR Scope						
2.	List the Contract, Case and/or PAR Deliverable Items						
3.	Prepare Project Breakdown Structure (PBS)						
4.	Identify Responsible and Performing Organization for Every Required Task						
5.	Correlate Project Chart of Accounts to PBS						
E.	*PROJECT MASTER PLAN, BUDGET AND SCHEDULE*						
1.	Identify and Define Project Level Milestones and Interfaces						
2.	Prepare the Project Master Plan and Schedule						

350

3. Prepare the Budget Section, Project Cost Performance Reports

4. Prepare the Project Detailed Network or Milestone Plan and Schedule

5. Identity the Critical Schedule Items

6. Establish Project Direct Labor and Other Source Requirements

7. Obtain Resource Availability Reports

8. Assure Project & Task Plans and Schedules are Based on Available Resources

F. *TASK PLANS, BUDGETS, AND SCHEDULES*

1. Prepare Task and Subtask Schedules and Budgets

2. Prepare Task and Subtask Plans and Status Reports

G. *WORK AUTHORIZATION AND CONTROL*

1. Issue Approved Task/Sub-task Budget and Work Order for Each Task or Subtask

2. Prepare Initial Task/Sub-task Plan and Status Report for Each Task or Sub-task

3. Issue Contracts or Purchase Orders for Outside Organizations

4. Release Shop Orders

5. Release Internal Work Authorization Documents

351

Figure A.1d   Project start-up checklist.

ITEM NO.	TASK DESCRIPTION	REQUIRED YES	REQUIRED NO	REQUIRED DATE	ASSIGNED RESPONSIBILITY	PR.	NOTES REFERENCE
H.	*PROJECT EVALUATION AND CONTROL*						
1.	Institute Financial Reporting System						
2.	Institute Manpower Reporting System						
3.	Institute Progress Reporting System						
	Initiate Weekly Updating of:						
4.	Task/Sub-task Plan and Status Reports						
5.	Task/Sub-task Labor Hour Variance Reports						
6.	Project Summary Material and Labor Variance Report						
	Initiate Monthly Updating of:						
7.	Project Master Plan and Schedule						
8.	Project Cost Performance Reports						
9.	Project Labor Hours Status Chart						
10.	Project Material Cost Status Chart						
11.	Initiate Periodic Project Evaluation & Review Meetings						
12.	Institute Action Assignment Procedures						
I.	*MANAGEMENT AND CUSTOMER REPORTING*						
1.	Establish Management Reporting Procedures						
2.	Initiate Submittal of Monthly Mgmt. Report on Project						
3.	Initial Submittal of Monthly Reports to Customer						

352

J. MARKETING AND CONTRACT ADMINISTRATION

1. Obtain Official Contract Documents
2. Define Contract Requirements
3. Clarify Type and Status of Contract
4. Establish Official Channels of Communication
5. Establish Personal Contacts with Customer
6. Arrange for Overtime Forecasts and Approval
7. Clarify Prime Booking Policy
8. Clarify Security Requirements
9. Determine Contract Close-Out Requirements
10. Establish Proof of Shipment Requirements
11. Provide for Negotiations with Customer
12. Define Waivers Granted
13. Assure That Accountability is Maintained
14. Define Deliverable Items
15. Clarify with Customer Acceptance Test Requirements
16. Institute a Project Public Relations Function

K. GOVERNMENT OR CUSTOMER FURNISHED EQUIPMENT & SERVICES (GFE)

1. Determine GFE Requirements
2. Determine Method of Submitting Requests to Customer
3. Establish Approval Requirements and Cycle
4. Identify Source of GFE
5. Issue Procedures to be Used in Obtained GFE
6. Clarify Method of Obtaining Funds for GFE
7. Establish Format and Procedures for Documentation
8. Institute Procedures for Negotiating Cost and Fee
9. Institute Procedures for Handling GFE Spares

Figure A.1e Project start-up checklist.

ITEM NO.	TASK DESCRIPTION	REQUIRED		REQUIRED DATE	ASSIGNED RESPONSIBILITY	PR.	NOTES REFERENCE
		YES	NO				
L.	*EXTENSIONS–NEW BUSINESS*						
1.	Assign Responsibility for System/Project Growth						
	Establish Processing Procedures For:						
2.	Contract Extensions and New Business						
3.	Changes in Scope to the Contract						
4.	Establish the System Growth Objectives						
5.	Prepare Booking and Business Forecasts						
6.	Prepare Program Brochures, Papers, Articles						
M.	*PROJECT RECORDS CONTROL*						
1.	Determine Records Control Requirements						
2.	Physical Equipment, Space and Personnel Required						
3.	Method of Control, Filing, Routing, etc.						
4.	Reproduction Methods and Requirements						
5.	Special Requirements for Classified Documents						
6.	Mail Distribution and Messenger Service						
7.	Central Facility and/or Satellite						
8.	Procedures for Special Requests						
9.	Procedures for Out-of-Plant Mail						
10.	Establish Standard Distributions						
11.	Maintain Significant Project Historical Data						
N.	*PURCHASING AND SUBCONTRACTING LIAISON AND POLICIES*						
1.	Determine Major Subcontractors and Vendors						

354

2. Establish Responsibility and Approval Cycles for Contractual Documents

3. Establish System to Monitor Subcontractor Cost, Manpower, Status, Progress, Problems, etc.

4. Establish Policy for Proposal Negotiation

5. Establish Policy for Changes in Scope

6. Provide Subcontractor Liaison

7. Comply with Small Business Requirements

8. Establish Make or Buy Policy

9. Develop Procedures for Subcontractor Close-Out for Each Subcontractor or Vendor,
   Determine:

10. Scope and Other Contractual Requirements

11. Contract Type and Status

12. Financial Status

13. Overtime Requirements and Approval

14. Security Requirements

15. Proof of Shipment Requirements

16. Procedures for Residual Inventory

17. Special Provisions

O. *SYSTEM ENGINEERING AND PLANNING*

1. Define Functional Requirements of System

2. Provide System Block Diagram

3. Establish Major Subsystems

4. Indicate Major Inputs and Outputs

5. Determine Performance Requirements for Major Performance Studies & Computer Simulations For:
   Accuracies

Figure A.1f   Project start-up checklist.

ITEM NO.	TASK DESCRIPTION	REQUIRED		REQUIRED DATE	ASSIGNED RESPONSIBILITY	PR.	NOTES REFERENCE
		YES	NO				
6.	Errors						
7.	Reliability & Redundancy Requirements						
8.	Deployment						
9.	Strategy						
10.	Environmental Factors (Natural & Artificial)						
11.	New Techniques						
12.	Man/Machine Considerations						
13.	Priorities						
14.	Trade-offs						
15.	Alternate Approaches						
16.	Establish System Growth Objectives						
17.	Establish Plan for System Improvement						
18.	Deliverable Computer Programs						
19.	Checkout/Exercising Equipment & Support						
20.	Data Extraction Areas & Data Reduction						
21.	System Integrity & Integration						
22.	Establish Technical Performance Reserves						
23.	Prepare Functional Cost Effectiveness Model						
P.	*PROJECT ENGINEERING*						
	Detail Equipment Specifications						
1.	Design						
2.	Performance						
3.	Environmental						
4.	Concept Approval						
5.	Design Approval						

	Interface (Input-Output)			
6.	Interface (Input-Output)			
7.	Drawing			
8.	Reliability			
9.	Test			
10.	Shakedown Spares			
11.	Site Support			
	Specifications			
12.	Overall System (Performance & Test)			
13.	Subsystem			
14.	Interface (Tie Subsystems Together)			
15.	Integration (Test Subsystems Together)			
	Test Facility			
16.	Single Cabinets			
17.	Single Subsystems			
18.	Interconnected Subsystems			
19.	Entire System			
20.	Preliminary Engineering Inspections (PEI)			
21.	Development Engineering Inspection (DEI)			
22.	Equipment Scale Models and Mock-ups			
23.	Establish Formal Design Freeze Requirements			

Q. *DESIGN AND DEVELOPMENT EQUIPMENT ENGINEERING*

1.	Prepare Technique Studies			
	Development and/or Design			
2.	Electrical			
3.	Mechanical			
4.	Breadboard			
5.	Initiate Drafting			

357

Figure A.1g Project start-up checklist.

ITEM NO.	TASK DESCRIPTION	REQUIRED		REQUIRED DATE	ASSIGNED RESPONSIBILITY	PR.	NOTES REFERENCE
		YES	NO				
6.	Write Vendor Specifications						
7.	Write Test Specifications						
8.	Release of Purchase Orders						
9.	Establish System for Drafting Follow-up						
10.	Establish System for Model Shop Follow-up						
11.	Establish System for Testing Equipment						
12.	Rework Equipment (ECN Writing)						
13.	Write Reports (Status, Technical, etc.)						
14.	Prepare and Review Inputs to Review Boards						
15.	Equipment Power Requirements						
16.	Test Equipment (Special and Standard)						
17.	Aerospace Ground Equipment						
R.	PROJECT INTEGRATION AND EQUIPMENT SUPPORT						
1.	Prepare Hardware, Material, Environmental Specs						
	Establish Design Engineering Requirements For:						
2.	Equipment Nameplates						
3.	Instructions to Packing						
4.	Equipment Ident. & Installation Location						
5.	Interface Wiring						
6.	Issue Joint Quality Control Document						
7.	Institute Configuration Control Program						
8.	Establish Design Review Program						
9.	Establish Design Change Control Procedure						

358

S. *MANUFACTURING POLICIES AND LIAISON*

1. Establish Method of Monitoring Factory
2. Determine Factory Release Requirements
3. Prepare Detailed Mfg. Release Plan
   Institute Factory Follow-up Procedures Covering:
4. Release by EN or ER (Eng. Notice or Release)
5. First Piece Review & Approval
6. Factory Engineering Problems, ECN's
7. Monitoring of Test Methods & Procedures

T. *RELIABILITY–PRODUCT ASSURANCE*

1. Establish Project Reliability Plan
2. Establish Method For Material Control
3. Prepare Product Assurance Contract Checklist
4. Prepare Quality Requirements for Eng. Docum.
   Monitor Project by Establishing:
5. System Reliability Models
6. Reliability Predictions
7. Reliability Failure Reporting System
8. Reliability Model Demonstration
9. Reliability Testing Procedure
10. Qualification Testing Procedure
11. Witnessing of Test by Quality Control
12. Reliability Allocations
13. Subcontractor Evaluation System
14. Quality Inspection System (Internal)
15. Gov't. (Customer) Inspection System (External)

Figure A.1h   Project start-up checklist.

359

ITEM NO.	TASK DESCRIPTION	REQUIRED		REQUIRED DATE	ASSIGNED RESPONSIBILITY	PR.	NOTES REFERENCE
		YES	NO				
U.	*ENGINEERING DOCUMENTATION*						
1.	Establish Eng' Rg. Documentation Req' Mts.						
2.	Prepare Mfg. Drawing Requirements						
3.	Tabulate System Assembly Drawings (SAD—Family Tree)						
4.	Prepare Functional Block Diagrams						
5.	Prepare Installation Drawings						
6.	Determine Drawing Grade for Instruction Books						
7.	Determine Customer Req'mts., Microfilm/Drawing						
8.	Accumulate, Tabulate Vendor Data						
	Establish EAM or EDP Program For:						
9.	Top Down Breakdown						
10.	Maintenance Parts List						
11.	Detail Parts Breakdown						
12.	Summary Equipment Listing						
	Implement System For:						
13.	Nomenclature Request						
14.	Federal Stock Numbers						
15.	Nameplate Approvals						
16.	Prepare Instruction Manuals						
17.	Maintain Complete Project Specification Docum.						
18.	Maintain Complete Project Report Documentation						
V.	*CONSTRUCTION AND EMPLACEMENT INSTALLA-TION*						
1.	Develop C&E Drawings & Specifications						

2. Review Customer Construction Plans and Specs
3. Special Studies to Establish Design Criteria
4. Obtain Siting & Configuration Requirements
5. Establish On-Site Customer Support
6. Develop Plan & Specs for Construction & Empl.
7. Prepare Emplacement Instruction Documents
8. Procure Material for Emplacement
9. Procure Service for Emplacement

W. SPARE PARTS

1. Establish Policy Agreement with Customer
2. Determine Type of Contract for Spares
3. Establish Method of Generating Requirements
4. Clarify Customer Approval Requirements
5. Decide Who Will Provide Spares
6. Determine Method of Funding
7. Establish Interim Release Provisions
8. Determine Method of Procurement
9. Determine Loadings (Requisition Engineering, Pack Transportation, etc.)
10. Write Procedures for Documenting & Negotiating
11. Determine Procedures for Changes to Requirements
12. Institute Requirements for Provisioning Documentation
13. Institute Requirements for Item Description Program
14. Identify Specifications Invoked

Figure A.1i Project start-up checklist.

361

ITEM NO.	TASK DESCRIPTION	REQUIRED YES	REQUIRED NO	REQUIRED DATE	ASSIGNED RESPONSIBILITY	PR.	NOTES REFERENCE
X.	*SITE TECHNICAL OPERATIONS*						
1.	Plan and Conduct Training Programs						
2.	Prepare Qualitative Personnel Requirements						
3.	Prepare Quantitative Personnel Requirements						
4.	Prepare Installation, Checkout & Test Plan						
5.	Prepare Operating & Maintenance Plan						
6.	Prepare Operating & Maintenance Documents						
7.	Manage Site Installation Activity						
8.	Perform System Checkout Acceptance Tests						
9.	Participate in Product Improvement and Modifications						
Y.	*SITE ADMINISTRATION OPERATIONS*						
1.	Select the Site Operations Manager & Staff						
2.	Institute Site Operating Procedures						
3.	Establish Administrative & Personnel Support						
	Establish General Services:						
4.	Mail, Message Center, Document Control, etc.						
5.	Financial, Cashier, T&E Control, etc.						
6.	Communications, TWX, Phone, Radio, etc.						
	Establish Technical Services:						
7.	Machine Shop, Model Shop						
8.	Test Equipment Laboratory Facilities						
9.	Drafting, Photographic, Weather						
	Establish Personal Services:						
10.	Housing, Food, Laundry						
11.	Medical, Dental, Mortuary, Religious						

12.	Protective, Safety, Security			
13.	Recreation, Company or Government Store			
14.	Prepare a Brochure on Site Accommodations			
Z.	TRANSPORTATION & LOGISTICS			
1.	Verification of Shipments and Receipts			
	Establish Passenger and Freight System			
2.	Commercial and/or Military			
3.	Domestic and/or Foreign			
4.	Establish with Customer POE and POD (Port of Embarkation and Delivery)			
5.	Institute Courier System, Courier Box			
6.	On-Site Transportation Requirements			
7.	Special Air Charters			
8.	Transportation Plan for Equipment			
9.	Warehousing, Staging, Material Handling			

Figure A.1j Project start-up checklist.

363

APPENDIX B

❮ ❯

Project Close-Out Checklists

As discussed in Chapter 16, project close-out is a very demanding assignment that requires considerable discipline to achieve a complete termination on the established date.

The following project close-out checklists are presented as an aid in planning and controlling the actions necessary to terminate the project quickly and efficiently. The benefits of using such checklists are that they:

- Clearly indicate the close-out functions and responsibilities, reducing ambiguity and uncertainty.
- Reduce oversight of important factors.
- Permit close-out progress to be monitored.
- Aid project team members with little or no experience in closing out a project.
- Inform other project team members about the activities of others during the close-out phase.

Figure B.1 contains 14 checklists, as indicated in its index. These are presented as a starting point for each organization to develop its own checklists reflecting the particular needs of its specific projects.

Index

A. Project Office (PO) and Project Team (PT) Organization
B. Instructions and Procedures
C. Financial
D. Project Definition
E. Plans, Budgets, and Schedules
F. Work Authorization and Control
G. Project Evaluation and Control
H. Management and Customer Reporting
I. Marketing and Contract Administration
J. Extensions—New Business
K. Project Records Control
L. Purchasing and Subcontracting Liaison and Policies
M. Engineering Documentation
N. Site Operations

Figure B.1a Project close-out checklists index.

PROJECT TITLE _____   COMPLETION DATE___

CONTRACT No. _____   COST TYPE _____

CUSTOMER _____   PROJECT MGR. _____

THE PROJECT CLOSE-OUT CHECK LISTS ARE DESIGNED FOR USE IN THE FOLLOWING MANNER:

COLUMN I—ITEM No.—Each task listed is identified by a specific number and grouped into categories. Categories are based on functions, not on organizations or equipment.

COLUMN II—TASK DESCRIPTION—Task descriptions are brief tasks that could apply to more than one category are listed only in the most appropriate category.

COLUMN III—REQUIRED, YES OR NO—Check whether the item listed applies to the project.

COLUMN IV—DATE REQUIRED—Insert the required date for accomplishment of the task.

COLUMN V—ASSIGNED RESPONSIBILITY—Insert the name of the man responsible to see that the task is accomplished on schedule. This may be a member of the Project Office or an individual within a functional department.

COLUMN VI—PRIORITY (PR)—A priority system established by the Project Manager may be used here, e.g., Priority # 1 may be all tasks that must be accomplished before the contractual completion date, Priority # 2 within 2 weeks after the completion date, etc.

COLUMN VII—NOTES, REFERENCE—Refer in this column to any applicable Procedures, a government specification that may apply to that task, etc.

Figure B.1b   Project close-out checklists instruction.

ITEM NO.	TASK DESCRIPTION	REQUIRED		REQUIRED DATE	ASSIGNED RESPONSIBILITY	PR.	NOTES REFERENCE
		YES	NO				
A.	*PROJECT OFFICE (PO) AND PROJECT TEAM (PT) ORGANIZATION*						
1.	Conduct project close-out meeting						
2.	Establish PO and PT release and reassignment plan						
3.	Carry out necessary personnel actions						
4.	Prepare personal performance evaluation on each PO and PT member						
B.	*INSTRUCTIONS AND PROCEDURES*						
	Issue Instructions for:						
1.	Termination of PO and PT						
2.	Close-out of all work orders and contracts						
3.	Termination of reporting procedures						
4.	Preparation of final report (s)						
5.	Completion and disposition of project file						
C.	*FINANCIAL*						
1.	Close out financial documents and records						
2.	Audit final charges and costs						
3.	Prepare final project financial report (s)						
4.	Collect receivables						
D.	*PROJECT DEFINITION*						
1.	Document final approved project scope						
2.	Prepare final project breakdown structure and enter into project file						

Figure B.1c Project close-out checklists.

367

ITEM NO.	TASK DESCRIPTION	REQUIRED YES	REQUIRED NO	REQUIRED DATE	ASSIGNED RESPONSIBILITY	PR.	NOTES REFERENCE
E.	*PLANS, BUDGETS, AND SCHEDULES*						
1.	Document actual delivery dates of all contractual deliverable end items						
2.	Document actual completion dates of all other contractual obligations						
3.	Prepare final project and task status reports						
F.	*WORK AUTHORIZATION AND CONTROL*						
1.	Close out all work orders and contracts						
G.	*PROJECT EVALUATION AND CONTROL*						
1.	Assure completion of all action assignments						
2.	Prepare final evaluation report (s)						
3.	Conduct final review meeting						
4.	Terminate financial, manpower, and progress reporting procedures						
H.	*MANAGEMENT AND CUSTOMER REPORTING*						
1.	Submit final report to customer						
2.	Submit final report to management						
I.	*MARKETING AND CONTRACT ADMINISTRATION*						
1.	Compile all final contract documents with revisions, waivers and related correspondance						

2.	Verify and document compliance with all contractual terms			
3.	Compile required proof of shipment and customer acceptance documents			
4.	Officially notify customer of contract completion			
5.	Initiate and pursue any claims against customer			
6.	Prepare and conduct defense against claims by customer			
7.	Initiate public relations announcements re contract completion			
8.	Prepare final contract status report			
J.	**EXTENSIONS–NEW BUSINESS**			
1.	Document possibilities for project or contract extensions, or other related new business			
2.	Obtain commitment for extension			
K.	**PROJECT RECORDS CONTROL**			
1.	Complete project file and transmit to designated manager			
2.	Dispose of other project records as required by established procedures			
L.	**PURCHASING AND SUBCONTRACTING**			
	For each Purchase Order and Subcontract:			
1.	Document compliance and completion			
2.	Verify final payment and proper accounting to project			
3.	Notify vendor/contractor of final completion			

Figure B.1d   Project close-out checklists.

369

ITEM NO.	TASK DESCRIPTION	REQUIRED		REQUIRED DATE	ASSIGNED RESPONSIBILITY	PR.	NOTES REFERENCE
		YES	NO				
M.	*ENGINEERING DOCUMENTATION*						
1.	Compile and store all engineering documentation						
2.	Prepare final technical report						
N.	*SITE OPERATIONS*						
1.	Close down site operations						
2.	Dispose of equipment and material.						

Figure B.1e Project close-out checklists.

370

Bibliography

Brent, J. A., and A. Thumann. *Project Management for Engineering and Construction.* Lilburn, GA: The Fairmont Press, 1989.

Briner, W., M. Geddes, and C. Hastings. *Project Leadership.* New York: Van Nostrand Reinhold, 1990.

Busch, D. H. *The New Critical Path Method.* Chicago: Probus, 1991.

Cleland, D. I., and H. Kerzner. *Engineering Team Management.* New York: Van Nostrand Reinhold, 1986.

Cleland, D. I., and W. R. King (Eds.). *Project Management Handbook* (2nd Ed.). New York: Van Nostrand Reinhold.

Dinsmore, P. C. *Human Factors in Project Management* (Rev. Ed.). New York: AMACOM, 1990.

Fleming, Q. W. *Cost / Schedule Control Systems Criteria.* Chicago: Probus, 1988.

Fleming, Q. W., and Q. J. Fleming. *Subcontract Project Management and Control.* Chicago: Probus, 1991.

Gareis, R. (Ed.) *Handbook of Management by Projects.* Vienna: MANZ, 1990.

Hastings, C., P. Bixby, and R. Chaudry-Lawton. *The Superteam Solution.* San Diego: University Associates, 1987.

Kezsbom, D. S., D. L. Schilling, and K. A. Edward. *Dynamic Project Management.* New York: Wiley, 1989.

Kharbanda, O. P., and E. A. Stallworthy. *Management Disasters and How to Prevent Them.* Brookfield, VT: Gower, 1986.

Kimmons, R. L., and J. H. Loweree (Eds.) *Project Management—A Reference for Professionals.* New York: Marcel Dekker, 1989.

371

Ludwig, E. E. *Applied Project Engineering and Management* (2nd Ed.) Houston: Gulf, 1988.

Morris, P. W. G., and G. H. Hough. *The Anatomy of Major Projects.* New York: Wiley, 1987.

Mulvaney, J. *Analysis Bar Charting—A Simplified Critical Path Technique.* Bethesda, MD: Management Planning and Control Systems, 1969.

O'Brien, J. J. *CPM In Construction Management* (3rd Ed.) New York: McGraw-Hill, 1984.

Rosenau, M. D. *Successful Project Management* (2nd Ed.) New York: Van Nostrand Reinhold, 1992.

Rosenau, M. M. *Faster New Product Development.* New York: AMACOM, 1990.

Whitten, N. *Managing Software Development Projects.* New York: Wiley, 1990.

Zells, L. *Managing Software Projects—Selecting and Using PC-Based Project Management Systems.* Wellesley, MA: QED Information Sciences, 1990.

Also see the references given for each chapter.

Notes

Chapter 1

1. Archibald, Russell D., and Alan Harpham, "Project Managers' Profiles and Certification Workshop Report," *Proceedings of the 14th International Expert Seminar,* March 15–17, 1990, the INTERNET International Project Management Association, Zurich, p. 8.

2. Adapted from Archibald, Russell D., "Implementing Business Strategies Through Projects," *Strategic Planning and Management Handbook,* William R. King and David I. Cleland, Eds. (New York: Van Nostrand Reinhold, 1987) Chapter 29, pp. 499–502.

3. Adapted from the testimony of Russell D. Archibald, "Project Management," on the Diablo Canyon Rate Case with the California Public Utilities Commission, Application Nos. 84-06-014 and 85-08-025, Exhibit No. 11,175, March 1987, pp. 26–28.

4. Testimony of Russell D. Archibald, *op. cit.,* pp. 50–52.

5. Adapted from Galbraith, John Kenneth, *The New Industrial State* (New York: The New American Library, 1968) pp. 25–28.

6. Toffler, Alvin, *Future Shock* (New York: Random House, 1970) p. 125.

7. *Ibid.,* p. 132.

8. *Ibid.,* p. 133.

Chapter 2

1. Gilbreath, Robert D., "Working with Pulses, Not Streams: Using Projects to Capture Opportunity," *Strategic Planning and Management Handbook,* William

R. King and David I. Cleland, Eds. (New York: Van Nostrand Reinhold Co., 1987) p. 6.

Chapter 3

1. Youker, Robert, "Organizational Alternatives for Project Management," *Project Management Quarterly, VIII*, No. 1, March 1975.

2. *Ibid.*

Chapter 4

1. Archibald, Russell D., and Alan Harpham, "Project Managers' Profiles and Certification Workshop Report," *Proceedings of the 14th International Expert Seminar*, March 15–17, 1990, the INTERNET International Project Management Association, Zurich, p. 9.

2. *Ibid.* This list is based on the work of Roland W. Spuhler.

3. Wilemon, David L., and Gary R. Gemmill, "Interpersonal Power in Temporary Management Systems," *Journal of Management Studies,* October 1971, pp. 319–320.

4. Archibald, *op. cit.,* p. 10.

Chapter 5

1. Briner, Wendy, Michael Geddes, and Colin Hastings, *Project Leadership* (New York: Van Nostrand Reinhold, 1990) pp. 41–46.

2. Hastings, Colin, Peter Bixby, and Rani Chaudhry-Lawton, *The Superteam Solution* (San Diego: University Associates, 1987) pp. 32–42.

3. *Ibid.,* pp. 35–37.

4. Owens, Stephen D., "Project Management and Behavioral Research Revisited," *Project Management Institute Proceedings* (Toronto 1982), p. II-F.1.

5. Briner, *op. cit.,* p. 18–30.

6. Thamhain, Hans J., and David L. Wilemon, "Conflict Management in Project Life Cycles," *Sloan Management Review,* Summer, 1975.

7. *Ibid.*

8. *Ibid.*

9. *Ibid.*

10. Burke, R. J., "Methods of Resolving Interpersonal Conflict," *Personnel Administration,* July–August, 1969, pp. 48–55.

11. Although Burke's study was conducted on general management personnel, it offers an interesting comparison to our research. Burke's paper notes that "Compromising" and "Forcing" were effective in 11.3% and 24.5% of the cases while "Withdrawal" or "Smoothing" approaches were found mostly ineffective in the environment under investigation.

12. Thamhain and Wilemon, *op. cit.*

13. Mower, Judith and David Wilemon, "A Framework for Developing High-Performing Technical Teams," *Engineering Team Management,* David I. Cleland and Harold Kerzner (New York: Van Nostrand Reinhold, 1986) p. 297–298.

Chapter 6

1. Elmes, M., and D. Wilemon, "Organization Culture and Project Leader Effectiveness," *Project Management Journal, XIX,* 4, 1988, pp. 54–63.

2. Lawrence, P., and J. Lorsch, "New Managerial Job: The Integrator," *Harvard Business Review,* November–December, 1967, pp. 142–151.

3. Buchanan, B., "To Walk an Extra Mile: The Whats, Whens, and Whys of Organizational Commitment," *Organizational Dynamics,* Spring, 1985.

4. Buchanan, B., "Building Organizational Commitment: The Socialization of Managers in Work Organizations," *Administrative Science Quarterly, 19,* 1974, pp. 533–546.

5. Mowdray, R., L. Porter, and R. Steers, *Employee-Organization Linkages: The Psychology of Commitment, Absenteeism, and Turnover* (New York: Academic Press, 1982).

6. Salancik, G., "Commitment and the Control of Organizational Behavior and Belief," in B. M. Staw and G. Salancik (eds.) *New Directions in Organizational Behavior* (Chicago: St. Clair, 1977).

7. Kouzes, J. M., and B. Z. Posner, *The Leadership Challenge* (San Francisco: Jossey-Bass, 1988) p. 108.

8. See note 1, p. 54.

9. Sathe, V., *Culture and Related Corporate Realities* (Homewood, IL: Richard D. Irwin, Inc., 1985) p. 12.

10. Winograd, T., and F. Flores, *Understanding Computers and Cognition* (Norwood, NJ: Ablex, 1986).

11. Ono, D., and R. Archibald, "Project Start-Up Workshops: Gateway to Project Success," *Proceedings of the Project Management Institute Seminar/Symposium,* Drexel Hill, PA: PMI, September 17–21, 1988, pp. 50–54.

12. Eden, D., "Self-Fulfilling Prophecy as a Management Tool: Harnessing Pygmalion," *Academy of Management Review, 9,* No. 1, 1984, pp. 64–73.

13. Wriston, W., *Risk and Other Four-Letter Words* (New York: Harper & Row, 1986).

Chapter 8

1. PERT: Program Evaluation and Review Techniques; CPM: Critical Path Method; PDM: Precedence Diagram Method. This topic is discussed in more detail in Chapter 10.

2. Boznak, Rudolph G., "Master Business Planning—The Art of Controlling Project Management in a Multi-Project Environment," *Proceedings of the Project Management Institute Seminar/Symposium,* Project Management Institute, Drexel Hill, PA, October 1987, pp. 143–158.

Chapter 10

1. Thamhain, Hans J., and David L. Wilemon, "Criteria for Controlling Projects According to Plan," *Project Management Journal,* Project Management Institute, Drexel Hill, PA, June 1986.

2. *"Project Management Body of Knowledge,"* Project Management Institute, Drexel Hill, PA, March 28, 1987, p. A-5.

3. *Ibid.,* pp. A-1 to H-7.

4. Nicholson, R. F., and E. M. Sieli, "Integrating Process Management With Project Management," *Proceedings of the Project Management Institute Seminar/ Symposium,* Project Management Institute, Drexel Hill, PA, October 1990, pp. 163–172.

5. Morris, Peter W. G., "Initiating Major Projects—The Unperceived Role of Project Management," *Proceedings of INTERNET '88,* Glasgow, 1988, The INTERNET-International Project Management Association, Zurich, p. 810.

6. Souder, William E., "Selecting Projects That Maximize Profits," Chapter 7, *Project Management Handbook,* 2nd ed., David I. Cleland and William R. King (eds.) (New York: Van Nostrand Reinhold, 1988) pp. 140–164.

7. Pilcher, Roy, *Appraisal and Control of Project Costs* (London: McGraw-Hill, 1973).

8. Lichtenberg, Steen, "Experiences from a New Logic in Project Management," *Dimensions of Project Management* (Heidelberg: Springer-Verlag, 1990) pp. 137–154.

9. Lichtenberg, Steen, and Russell D. Archibald, "Experiences Using Next Generation Management Practices," *Proceedings of INTERNET '92,* Florence, Italy, INTERNET-International Project Management Association, Zurich.

10. Anon., *Military Standard Work Breakdown Structure for Defense Material Items,* MIL-STD 881A, U.S. Department of Defense, Government Printing Office, Washington, DC, April 25, 1975.

11. The Roseta Stone is a basalt tablet inscribed with a decree of Ptolemy in 196BC in Greek, Egyptian hieroglyphic and demonic; discovered in 1799 near the town of Roseta, Egypt, and considered by most scholars as the key to deciphering the ancient Egyptian hieroglyphics.

12. Anon., *Military Standard, Defense System Software Development.* DOD-STD-2167A, U.S. Department of Defense, Feb. 29, 1988.

13. Anon., *Work Breakdown Structure Guide,* U.S. Department of Energy, DE87-007606, Feb. 1987.

14. Cleland, David I., and William R. King, "Linear Responsibility Charts in Project Management," *Project Management Handbook,* 2nd Ed., David I. Cleland and William R. King (Eds.) (New York: Van Nostrand Reinhold, 1988) pp. 374–393.

15. Mulvaney, John, *Analysis Bar Charting, A Simplified Critical Path Analysis Technique,* Management Planning and Control Systems, Bethesda, MD, 1969.

16. O'Brien, James J., *CPM In Construction Management,* 3rd Ed. (New York: McGraw-Hill, 1984).

17. Parkinson, C. N., *Parkinson's Law* (Boston: Houghton Mifflin, 1957).

Chapter 11

1. *INTERNET Handbook of Project Start-Up,* Fangel, M. (ed.), INTERNET International Project Management Association, Committee on Project Start-Up, Saettedammen 4, DK-3400, Hilleroed, Denmark, 1989.

2. Pincus, Claudio, "A Workshop Approach to Project Execution Planning," *Project Management, A Reference for Professionals,* Robert L. Kimmons and James H. Loweree, Eds. (New York: Marcel Dekker, 1989) pp. 349–355.

3. Pincus, *op. cit.,* p. 351.

4. Gillis, Robert, "Strategies for Successful Project Implementation," INTERNET *Handbook of Project Start-Up, op. cit.,* Section 4.

5. Ono, Dan, and Russell D. Archibald, "Project Start-Up Workshops: Gateway to Project Success," *Proceedings of the Project Management Institute Seminar / Symposium,* Project Management Institute, Drexel Hill, PA, Sept. 17–21, 1988, pp. 50–54.

Chapter 12

1. Kemps, Robert R., "Solving the Baseline Dilemma," *Performance Management Association (PMA) Newsletter,* Performance Management Association, 101 S. Whiting St., Suite 313, Alexandria, VA 22304, Autumn, 1990.

2. Fleming, Quentin W., *Cost / Schedule Control Systems Criteria, The Management Guide to C / SCSC* (Chicago: Probus Publishing Co., 1988) pp. 122–124.

3. Fleming, *op. cit.*

Chapter 13

1. Adapted from Archibald, Russell D., "Project Interface Management: A Key to More Effective Project Management," *Proceedings of the 9th INTERNET*

World Congress on Project Management, Sept. 4–9, 1988, Glasgow, The INTERNET Project Management Association, Zurich, pp. 51–59.

2. Tuman, John Jr., "Development and Implementation of Project Management Systems," *Project Management Handbook,* 2nd Ed., David I. Cleland and William R. King (Eds.) (New York: Van Nostrand Reinhold, 1988) p. 666. Used by permission.

Chapter 14

1. Cleland, David I., "Defining a Project Management System," *Project Management Quarterly,* Project Management Institute, Drexel Hill, PA, December 1977.

2. Tuman, John Jr., "Development and Implementation of Project Management Systems," *Project Management Handbook,* 2nd Ed., David I. Cleland and William R. King (Eds.) (New York: Van Nostrand Reinhold, 1988) Chapter 27, pp. 652–691.

3. Levine, Harvey A., *Project Management Using Microcomputers* (Berkeley, CA: Osborne McGraw-Hill, 1986) p. xii.

4. Rosenau, Milton D. Jr., *Successful Project Management,* 2nd Ed. (New York: Van Nostrand Reinhold, 1992).

5. Levine, Harvey A., *op. cit.*

6. Lowery, Gwen, *Managing Projects with Microsoft Project for Windows* (New York: Van Nostrand Reinhold, 1990).

7. Halcomb, James, *Planning Big with MacProject* (Berkeley, CA: Osborne McGraw-Hill, 1986).

8. Yahdav, Daniel, *PM Solutions* (San Rafael, CA: 1 Soft Decision, Inc., 1991).

Chapter 15

1. Ono, Daniel P., and Russell D. Archibald, "Achieving Quality Teamwork Through Project Performance Reviews," *Proceedings of the Project Management Institute Annual Seminar/Symposium,* Project Management Institute, Drexel Hill, PA, Dallas, 1991.

Index

Action assignments, 333
Actual cost of work performed (ACWP), 228
Activities, conflict between, 141
  definition of, 215
Ad-hocracy, 22
Apportioned effort, 207
Authority, project, 52, 77, 80
  earned, 80
  effect of staffing methods on, 52
  legal, 80
  types of power related to, 80
Authorization documents, project, 258

Budget
  baseline, 266
  direct project, 222
  indirect project, 222
  task, 225
  total project, 221
  variance, 231, 279
  versus expenditures, 275
Budgeted cost of work performed (BCWP), 228, 279
Budgeted cost of work scheduled (BCWS), 228, 279

Change Control Board, 186, 270
Chart of accounts, project, 224
Close-out, 163, 339–343
  checklists, 340
  plan and schedule, 340
Commitment, 109–126
Computer-supported project management information systems, 302
Configuration management, 270
Conflict, 99
  handling modes, 102
  intensity over life cycle, 99
  management of, 105
  prediction of, 139
  recommendations for minimizing consequences, 105
  responsibility, 103
  sources of, 99
Contract, administration, 157, 160, 167, 270, 341
  control, 270
  release, 258
Control, of changes, 268
  of commitments, 275
  of contract, 270
  of cost, 274

379

Control *(Continued)*
  documents, 300
  of engineering changes, 270
  of products, 186
  of projects, 5, 178, 185, 257, 296, 330
  of project scope, 268
  requirements for, 271
  of schedule, 271
  of task work orders, 269
  of subcontracts, 270
  of work, 258
Correlation of schedule, costs and technical
  progress, 271
Cost, accounting problems, 277
  causes of problems, related to, 276
  to complete, 276
  at completion, 276
  control, 274
  of project management, 15
  performance reports, 280–287
  schedule performance reports, 280
  variance, 231, 279
Cost/Schedule Control Systems Criteria
  (C/SCSC), 228, 278
  cost/schedule performance reports, 280
  software packages for, 302–319
Critical path method (CPM), 142, 215, 220,
  231, 273
Customer reports, 334, 338

Dates, types of event, 211
Design reviews, 332
Development, projects, 35
  tasks, 226

Earned value, 277
Engineering change control, 270
Environment, project, 38
Environmental, actors and factors, 39–41
  linkages with project, 41
  scan, 39, 40
Evaluation, post-completion, 342
  project, 320–338
  review meetings, 328
Event(s), dates, 211
  definition of, 211
  identification of, 212
  interface, control of, 274
  interface and milestone checklist, 213
Expenditures, assurance of proper, 275
  measured against budget, 275
Extensions, project, 341

Field project manager duties, 176
Field operations task schedules and
  budgets, 228
File, project, 232
Functional manager, 78
Functional project leader, 5, 75, 78, 163

General Manager, 5, 73, 76, 85
Graphic displays, 323
Growth:
  projects, 10
  strategically managed, 13

Imperatives of technology, 14
Improvement, management development,
  17–21
  in organization of responsibilities, 19
  in project management, 16
  in systems, methods and procedures, 19
Input/output analysis, 292
Integrated project evaluation, 320
  schedule and cost control, 277
Integrative planning and control, 5, 179
  responsibilities, 5, 73–90
  roles, 72, 129
Interdepencies between and within
  projects, 142
Interface(s), documentation and control,
  294, 296
  event control, 274
  event definition, 211
  identification, 211, 294
  management, 53, 288, 294
  product, 291
  project, 291, 293
  scheduling, 295
  types of, 292
Interpersonal influence bases, 80
Inventory of projects, 20, 335

Late charges, 275
Life cycle of projects, 25–30
  conflict related to, 99
Leadership and commitment, 110
Level of effort, 207
Limited resource re-scheduling, 143

Management, configuration, 270
  development and training, 18
  interface, 53, 288, 294
  release, 259
  reports, 335

reserves, 223
  control of, 276
reserve transaction register, 276
Manager of projects, 48, 74, 84, 85
Manufacturing coordinator, 158, 175
Master contract release, 258
Master schedule, 213
Matrix, management, 44, 109
  task responsibility, 61, 93, 208
Measuring progress against schedule, 273
Measuring expenditures against budget,
  275
Milestone identification, 212
Multiple project(s), 38, 136–149
  central planning for, 57, 144
  large versus small, 138
  management, 38, 136
    objectives, 137
    operations planning and control, 57, 144
  manager, 48, 74, 84, 85
  master business planning, 147

Network, 142, 215, 220, 231, 273, 307
  project plan, 215
  systems, 142
  task plan, 231
  time analysis, 220
Nonrecurring costs, manufacturing tasks,
  226, 230

Objectives, hierarchy of, 7
  project, 93, 181
  management, 4
Operations planning and control, 57, 144
Organization, charting relationships, 60
  growth, 6
  improvements in, 19
  impact of project management on, 21
  location of program and project
    managers, 47
  matrix, 44, 109
  of program/project management
    function, 43
  of project office, 49, 154
  of project participants, 50
  of project support services, 53
  of project team, 49, 158
  responsibility matrix, 93, 208–210

PERT, (see program evaluation and reviews
  technique)
PDM, (see precedence diagram method)

Plan:
  baseline, 266
  close out, 340
  preinvestment, 186
  project funding, 224
  project summary, 185
  proposal, 186
Planning, documents, 178–235, 300
  functions and tools, 185–188
  project, 178–235
  and project life cycle, 179
  steps, 235
Post-completion evaluation, 342
Precedent diagram method (PDM), 142,
  215, 220, 231, 273, 307
Priorities, factors influencing, 140
  of projects, 139
  review board, 140
  rules, 141
Procurement task schedules and budgets,
  228
Product, planning and control functions
  and tools, 186
  design and development, 154
  installation, testing, and field support, 156
  manufacture, 155
  planning reviews, 332
  support services, 54
Program, definition of, 24
Program evaluation and review technique
  (PERT), 142, 215, 220, 231, 273
Progress measurement, 273, 275, 277, 283
Project(s), acceleration, cost of, 31
  accountant, 158, 174
  authority, 52, 80
  authorization, 258
  budget, 221, 222, 231, 258, 275, 279
  capital facilities, 35
  categories of, 33–37
  chart of accounts, 224
  close-out, 163, 339–343
  close-out checklists, 340
  commercial, 34
  compared to departments, 31
  construction, 35
  contract administrator, 157, 167–171
  control, 5, 178, 185, 257, 296, 330
  control center, 330
  controller, 158, 171–174
  critical review, 337
  development, 35
  defining the, 193–201

Project(s) *(Continued)*
  definition of, 24–37
  direction, 332
  directives, 333
  engineer, 157, 164
  environment, 38
  evaluation, 320–338
    methods and practices, 321
  extensions, 341
  field manager, 176
  file, 232
  functional leader, 5, 75, 78, 163
  funding plan, 224
  government, 34
  growth, 10
  information system, 36
  interfaces, 53, 211, 274, 288, 292–296
  interdependencies, 142
  life cycle, 25–30, 99
  management, 37
  multiple, 38, 136–149
  manufacturing coordinator, 158, 175
  master schedule, 213
  network plan, 142, 215, 220, 231, 273
  new product development, 35
  objectives, 93, 181
  office, 49, 154
  performance reviews, 329
  planning and control:
    functions and tools, 185–188
    integrated and predictive system, 5
    project manager responsibilities, 159,
      178
  planning, 178–235
  planning steps, 235
  priorities, 139
  as a process, 183
  register, 20, 335
  release, 258, 262
  research, product development and
    engineering, 35
  resource plans, 221
  review meetings, 328
  scope, 182, 268
  sponsor, 5, 73, 76
  staffing of, 49, 157
  stakeholders, 93
  start-up, 236, 238
    checklists, 238, 345–363
  summary plan, 185
  taskforce, 45
  team, 6, 49, 92, 107, 109, 154, 158

  team planning, 236–244
    need for, 237
    process, 238
Project breakdown structure, 93, 193, 240,
    271
  examples of, 197–206
  control of, 202
  cost performance report, 280
  creating, 195–196
  description of, 193
  integration with functional organizations,
      260–261
  use of, 201
Project evaluation, 320–337
  review meetings, 328
Project management, advantages of, 15
  cost of, 15
  failures in, 4
  forces behind, 14
  function, organizing the, 43–71
  importance of, 4
  improvements in, 16
  information systems (PMIS), 299–319
    definition, 299
    documents, 300
    important factors in usage of, 316
    software market trends, 304
    software selection, 310
  linked to strategic management, 9
  manager of, 48, 74
  matrix, 44, 109
  multiproject, 38, 136
  objectives, 4
  opportunities for improvement, 16
  organizational impact of, 21
    alternatives, 44
  overcoming barriers to, 127–135
  support services, 53
  symptoms of poor performance in, 17
  triad of key concepts, 4, 72, 91
Project manager, 5, 12, 156
  alternate ways of filling the role of, 81
  authority, 52, 77, 80
  career development, 89
  career planning, 89
  continuity of assignment, 82
  division of responsibilities, 84
  duties, 159–163
  field, 158
  full-time vs. part-time, 82
  influence bases, 80
  as interface manager, 289